SpringerBriefs in Geography

More information about this series at http://www.springer.com/series/10050

Deborah F. Shmueli • Rassem Khamaisi

Israel's Invisible Negev Bedouin

Issues of Land and Spatial Planning

 Springer

Deborah F. Shmueli
Department of Geography and
    Environmental Studies
University of Haifa
Haifa, Israel

Rassem Khamaisi
Department of Geography and
    Environmental Studies
University of Haifa
Haifa, Israel

ISSN 2211-4165   ISSN 2211-4173   (electronic)
SpringerBriefs in Geography
ISBN 978-3-319-16819-7   ISBN 978-3-319-16820-3   (eBook)
DOI 10.1007/978-3-319-16820-3

Library of Congress Control Number: 2015935741

Springer Cham Heidelberg New York Dordrecht London

Springer International Publishing AG Switzerland is part of Springer Science+Business Media
(www.springer.com)

This book is dedicated to our fathers:
Saul B. Cohen who is a constant source
of encouragement and insight, asking
the probing and difficult questions and
urging publication; in memory of Avshalom
Shmueli who lived, worked, and held deep
friendships with Negev Bedouin and from
whose early publications we were able
to learn historical sequence; and in memory
of Mohyee Alden Khamaisi, who as an
indigenous person belonged to the land,
made his way through life with courage
and generosity, and taught his family
to make their own path through life
as self-made pioneers in their homeland
to the benefit of humanity, and their families.

Acknowledgements

Over the years, there are many people in the Negev with whom we have met and worked who are an inspiration to us and to the writing of this book. Jabar Abu Kaf, Dr. Mohamad Al-Nabare, Atia Ala'sam, Ishak Abu Hmad, and Amram Kalagi are a few of the very special individuals whose continuous efforts and outlook may pave the way to resolution. The production of this book has benefitted greatly from the dedication and graphic skills of Noga Yoselevich.

Contents

About the Authors

Deborah F. Shmueli is a faculty member in the Department of Geography and Environmental Studies at the University of Haifa. She is a planner specializing in public and environmental policy issues related to water, land use, and public infrastructure. A major focus of her efforts is in environmental and public sector conflict management and community and institutional capacity building. Examples include targeting consensus-based conflict management capacity within communities, NGOs, local governments, and district and national Ministries. She has published numerous articles related to planning, conflict management and justice issues. She received her Ph.D. in Architecture and Urban Planning (1992) from the Technion, Israel Institute of Technology, and her M.C.P. and B.S. in Urban Planning from the Massachusetts Institute of Technology (MIT, 1980).

Rassem Khamaisi is a faculty member in the Department of Geography and Environmental Studies at the University of Haifa. He is an urban and regional planner and geographer, specializing in urban and rural geography, public administration, and urban management. A major focus of his efforts is on planning among the Arabs in Israel and the Palestinians in the Palestinian territory and Jerusalem. He publishes research in three languages—English, Hebrew, and Arabic. He received his PhD from the Hebrew University of Jerusalem (1993), M.Sc. in Urban Planning from the Technion, Israel Institute of Technology (1985), and his B.A. in Geography from the Ben Gurion University of the Negev (1981).

Professors Khamaisi and Shmueli have been studying and working with the Bedouin in various capacities, both assisting the Bedouin with planning processes and as members of various government planning teams, committees, and commissions whose land use policies impact Bedouin settlement.

Chapter 1
Introduction

Introduction

My name is Abed and I am a member of the Tiaha Bedouin tribe, the Ala'sam clan, living in the now recognized village of Abu Tlool in Israel's Naqab/Negev[1] desert. I live in the home which I inherited from my grandfather. This is our land, as is recognized by my relatives and all of the other tribes. But the State of Israel, of which we are citizens, does not. What does this mean for my family? We cannot improve our home and its land. The Ottoman Regime recognized our ownership of this land, giving my grandfather the land title ('Kosan Tabo'). Since then we have continued to live here, first in a tent, then in a tin hut and now in a cement house of about 100 square meters which we built. We live here with our eight children, without electricity, running water or paved roads.

I have served this country for years as a scout in the army but today I was in court to defend myself against a criminal offense. What is this crime? It is having built my house without having been issued a building permit, since state regulations had forced me to do so. I had asked the District Engineer to secure a building permit for me in accordance with the law. Without such a permit my home is scheduled for demolition. The engineer replied that he could not issue a building permit. When I asked why, he answered that there were two main reasons: the first is that I had built on state land which is still zoned as agricultural according to the Regional Plan; and secondly there is no authorized local outline plan which permits the issuance of a building permit. So I live under the constant threat of having my house demolished. My family and I are caught in a vicious circle as we try to survive in the face of many restrictions. We are not alone—there are thousands of Bedouin families such as mine here in the Negev who are waiting for solutions. Many government representatives have approached us with promises—none of which are fulfilled. My family and I no longer have faith in the government. My children do not see themselves as serving the country, as I have done, since it doesn't recognize our rights.

---

[1] In order to be consistent with the majority of academic literature and media coverage on the subject, we use the Hebrew term "Negev" in this chapter. The Negev is the semiarid southernmost region of Israel, comprising approximately 60 % of the country's area.

© Springer International Publishing Switzerland 2015
D.F. Shmueli, R. Khamaisi, *Israel's Invisible Negev Bedouin*,
SpringerBriefs in Geography, DOI 10.1007/978-3-319-16820-3_1

This is one of many such stories which we have heard from various Bedouin whom we have met during our years of work in the Negev. Some of them, like Abed, continue to live on their lands; others, forced to leave their lands in the early 1950s by military order, were resettled in the place where they now reside. Some of these hamlets are recognized, some in the process of recognition, and others remain unrecognized. Even those villages which were granted government recognition still have no permanent housing or infrastructure. Some Negev Bedouin have undergone further resettlement. The narratives of these people, our field observations, and participation in planning and municipal committees, in addition to studying governmental plans and reports and other pertinent literature, are the basis for writing this book.

Conflict roils around recognition, planning, and appropriate municipal frameworks for Israel's 224,200 Negev Bedouin (Central Bureau of Statistics (CBS) 2014, Table 2.16, page 131), especially the 55,700 Bedouin living outside localities in dispersed unauthorized villages (CBS 2014, Table 2.21, page 152), in tents and cinder-block shacks, accessible via dirt tracks, and lacking municipal water, sewage, or electricity. Today one of the critical problems facing the government involving the Arab community in Israel surrounds the rights of the Bedouin to recognition, housing, infrastructure solutions, and economic opportunities in the country's Southern District Negev (Hebrew) or Naqab (Arabic) desert.

The Bedouin are not a monolithic society, but, due to assumptions based on historic behaviors, are often treated en bloc discriminatorily by government agencies, and other Arab and Jewish communities. This has a direct impact on their low socioeconomic status. The reality is multifaceted. This book addresses the complexity of Bedouin issues, particularly as they relate to spatial planning, municipal organization, land regime, and ownership disputes.

The struggle for recognition of land claims by Israel's Bedouin has both universal and unique characteristics. The universality lies in the aspirations as an indigenous people to maintain cultures, traditions, and deeply rooted ways of life within a distinct physical setting and to gain land rights. It is unique in that the Bedouin are not geographically isolated in remote parts of a country, with little contact with the dominant society. Instead, the Bedouin territory in the Negev is enclosed by the adjoining, highly modernized Israeli Jewish society. The Bedouin are forced by governmental authorities and economic realities to abandon their seminomadic for a semi-urban way of life, which upon examination can be characterized as false or pseudo-urbanization. The national-religious affiliations, cultural and political behaviors, and demographic and geographical characteristics, compounded with the internal Bedouin communities' fragmentation, economic marginalization, and geopolitical conflict have resulted in the creation of unrecognized villages, building without permits, absence of basic municipal services, and lack of economic development.

Absent access to expanded grazing lands for their herds, the Negev Bedouin are locked in dispute with the Government of Israel over the small portion of the arid Negev into which they have been relocated by government fiat, and over control of their own local governance, including a voice in the planning process which regulates how they can use the lands on which they live. The position of the Israeli Government is influenced by military strategic concerns, as well as by its reluctance to undermine its centrally controlled local governance and land planning structures by ceding such control to a minority indigenous group. This reluctance is reinforced by the fear that recognizing the Bedouin land claims might be used as a precedent for such claims by the Israel's Arab populace which constitutes 21 % of the country's population.

Israel as a whole has a very complex social, cultural, and political fabric with territorial uncertainties. In the Negev context, the urgent need is to develop the Bedouin ties with the broader Israeli society to insure individual and collective stability and change both the perception and the reality of urbanization.

Other stakeholders involved include the Israeli Government and its various ministries (Defense, Interior, Housing, Infrastructure, and Environment) at the national, district, and local levels. Here too, there are diverse policy objectives both across and within ministries; the overriding perception is of the Bedouin as a growing demographic threat, as smugglers and a potential fifth column. The judiciary has participated, often at odds, with government decisions. Customary Bedouin versus Israeli State laws are also in conflict. The Jewish population at both the national and Southern regional levels have diverse views, some of which are based on misinformation.

These issues are approached from the standpoint of the various stakeholders and a flexible set of planning options for creating new, revised, and supportive municipal structures which might reconcile the various interests that are explored. This volume investigates the interface between land rights; administrative, judicial, and boundary decisions; and spatial planning and planning rights, within the contradictory contexts of state-imposed top-down policies and the Bedouin community's bottom-up counter-responses to these policies. The theoretical prism is one of indigenous land rights and counter-planning, urbanization in non-Western societies, and environmental/distributional justice.

The book is organized into nine chapters. Following this introduction, Chap. 2 provides definitions and introduces the Bedouin context, both historical and current; Chap. 3 provides an overview of the Arab minority as a whole in Israel, focusing on its status within Israeli society, spatial distribution, urbanization patterns and trends, and the Bedouin within this context. Chapter 4 sets forth the theoretical contexts of indigenous land claims, accelerated urbanization, and justice. Chapter 5 focuses on an overview of the Negev Bedouin situation, including a historic summary of land claims and the management of geographic territory. Chapter 6 provides an overview of the evolution of local authorities as a wider context for understanding Chapter 7 in which we zoom in on resettlement planning, land claims, establishment of urban communities and regional councils, and unrecognized villages, providing a synopsis of government policies and Bedouin responses. Chapter 8 details the lessons learned from past failed policies and the concluding chapter presents our proposals for a flexible Bedouin resettlement and collaborative planning strategy.

The Bedouin are diversified according to sociocultural affiliations, the degrees to which they experience rapid urbanization processes, and their land ownership status.

Differentiations also stem from the pull between traditional and modern lifestyles, tribal (clan affiliation) and individualism, rural and urban existences, and national–religious and citizenship affiliations. The Bedouin are a minority within a larger Arab minority, living in a peripheral geographic area which lacks the amenities of the more central areas of the country. They have a strong attachment to place and land, and resist government resettlement policies, due in most part to lack of trust based on previous experience.

In the context of the Bedouin of the Negev, the structural and behavioral contradictions between and among governmental agencies and community subgroups with regard to spatial planning and municipal reorganization may shed light on disputes involving other communities in geographical peripheries who are economically marginalized, undergoing cultural changes and crises, and politically weak.

The Bedouin live in various Arab countries in nomadic, seminomadic, and sedentary villages in the periphery, but their situation in Israel is unique along two main dimensions. The first is related to the Israeli land regime and spatial and territorial planning, which is directly affected by ideological and geopolitical considerations. The second relates to the individual and cultural divides—among state, customary, and religious laws and behaviors. It is hoped that the discussion, focusing on these dimensions, will provide not only an academic contribution but also a practical one.

The book's focus is the Bedouin of the Negev. Some components are unique to Israel and its relationships with its Arab citizens in general and the Bedouin in particular. However, the complexity of the situation has universal value regarding spatial planning and resettlement in peripheral areas for marginal communities existing amidst geopolitical and sociocultural conflict. We hope that the insights gained concerning complex interactions between states and their indigenous populations, issues of justice, forms of urbanization, and land claims, will be applicable to other situations around the world.

Chapter 2
Bedouin: Evolving Meanings

The term Bedouin is connected with seasonal nomadic behavior in arid deserts. Other definitions are associated with the terms meaning 'the beginning' (*al-Badia or Badia*), alluding to the Bedouin being original or indigenous. Today, most of the Bedouin in the Negev live in sedentary dwellings; two-thirds dwell in towns (2014), and less than 5 % work in agriculture or grazing. The Bedouin are Arab and Muslim, yet differentiate themselves from the larger Arab minority, affiliating themselves specifically as Bedouin. The origin of this term, its meaning, the Bedouin context, and consequences are the focus of this chapter.

Between Ethnicity and Lifestyle

The term Bedouin[1] in general and in the Israeli context in particular require reexamination in order to describe the new reality of this community. The term refers to "a nomadic Arab of the desert," describing the Arabs of the early seventh century

---

[1] The **Bedouin** (/'bɛduɪn/, also **Bedouins**; from the Arabic *badw* بَدْو or *badawiyyīn/badawiyyūn* بَدَويُّون, plurals of *badawī* بَدَوي) are a part of a predominantly desert-dwelling Arabian ethnic group traditionally divided into tribes or clans, known in Arabic as *'ashā'ir* (عَشائِر). The Bedouin form a part of, but are not synonymous with, the modern concept of Arabs. Bedouins have been referred to by various names throughout history, including Qedarites in the Old Testament and "Arab" by the Assyrians (*ar-ba-a-a* being a nisba of the noun *arab*, a name still used for Bedouins today). While most Bedouins have abandoned their nomadic and tribal traditions for a modern urban lifestyle, they retain traditional Bedouin culture with concepts of belonging to *'ašā'ir*, traditional music, poetry, dances (like Saas), and many other cultural practices. Urbanized Bedouins also traditionally organize cultural festivals, usually held several times a year, in which they gather with other Bedouins to partake in, and learn about, various Bedouin traditions—from poetry recitation and traditional sword dances to classes teaching traditional tent knitting and playing traditional Bedouin musical instruments. Traditions like camel riding and camping in the deserts are also

© Springer International Publishing Switzerland 2015
D.F. Shmueli, R. Khamaisi, *Israel's Invisible Negev Bedouin*,
SpringerBriefs in Geography, DOI 10.1007/978-3-319-16820-3_2

who were "Bedouin or desert nomads." It is derived from the old French *beduin*, based on the Arabic *badawī*, (plural) *badawīn*—"dwellers in the desert," from *badw* "desert" (http://www.oxforddictionaries.com/definition/americanenglish/Bedouin). The Encyclopaedia Britannica defines Bedouin as Arabic-speaking nomadic peoples of the Middle Eastern deserts of the Arabian Peninsula, Egypt, Israel, Iraq, Syria, Jordan, and North Africa (http://www.britannica.com/EBchecked/topic/58173/Bedouin). In Arabic the word *badawī*, is connected with the *Badia*, which is taken from the word beginning or original. Thus the Bedouin feel and define themselves as the original native community. They are hierarchically ordered as *badawī* (nomads), *falah/humran* (peasants), *abbed/sumran* (slaves or servants), rural, and urban. This traditional classification needs reframing in light of the rapid urbanization leading to the economic and functional transformation of Arab countries in the Middle East as well as the emergence of Israel as a nation state. The outcome has transformed the settlement networks and human behavior of the Bedouin. To say today that the populations of Riad or Jada in Saudi Arabia, Dubai or Kuwait in the Gulf countries, or even Maan in Jordan are Bedouin is not an apt description of reality, even though these groups try to maintain their identity as Bedouin, fueled by a nostalgic longing for the past.

The Bedouin currently represent only a small portion of the total population of Arab countries. In 2013, 21.2 million or 5.7 % of the total Arab population of 370 million inhabitants in Arab countries were classified as Bedouin. Despite these relatively small numbers, they inhabit or utilize a large part of the land area. Most of them are animal herders who migrate into the desert during the rainy winter season and move back toward the cultivated land in the dry summer months. Although Bedouin, as a matter of caste, traditionally despised agricultural work and other manual labors, many of them have become sedentary as a result of political and economic developments since World War II. In the 1950s, Saudi Arabia and Syria nationalized Bedouin rangelands, and Jordan severely limited goat grazing. Conflicts over land use between Bedouin herders on the one hand and settled agriculturists on the other have increased (http://www.britannica.com/EBchecked/topic/58173/Bedouin). The past three decades have witnessed the growth of large urban centers within Arab countries. This growth, catalyzed by national and regional plans and the provision of public services, includes Bedouin immigrants to these urban centers.

In the Israeli context, Bedouin does not describe a way of life of a nomadic community, but defines an ethnic group which is part of the Arab minority. Israeli policies do not deal with the Arab community as a national and collective minority; instead, they deal with sub-ethnic groups according to both religious and cultural affiliation (Muslims, Christians, Bedouin, and Druze) and geographic location (North-Galilee, Center-Triangle, and South-Negev) as well as according to the local governmental structures within which they reside. Such structures include Arab citizens living in heterogeneous cities, Arab municipalities, or Arab villages within

---

popular leisure activities for urbanized Bedouins who live within close proximity to deserts or other wilderness areas. http://en.wikipedia.org/wiki/Bedouin

regional councils. The Bedouin of the Galilee are treated as a separate community, in contrast to the Bedouin of the Negev (Israel's southern desert). Most of the Bedouin in the Negev accept and even prefer to identify themselves as Bedouin, despite their transformation from herders to a more sedentary lifestyle, some of them living in towns which offer private and public services, and in recognized and non-recognized villages.

The Bedouin affiliation is a sociocultural and kinship belonging to tribal traditional structures, as well as territorial and land ownership links; the latter are the focus of major dispute between the Negev Bedouin and the State of Israel. The strength of tribal kinship among the Bedouin of the Negev is a result of the following:

• Sociocultural and historical tribal affiliation: The Bedouin affiliate themselves with eight tribes (Azazmah, Tarabin, Tiaha, Hanagrah, Gubarat, Saidin, Ahewat, and Gahalin), consisting of about 95 sub-tribes or clans (this book will use the term clan), each occupying a territorial area (*H'ema*, protected and belonging to a certain tribe) (Fig. 2.1). Three of these, the Azazmah, Tarabin, and Tiaha, are the major tribes that remained in the Negev after the establishment of the Israeli State in 1948. Each aspires to continue to live on its historical territory. With Statehood in 1948, Israel resettled the Bedouin who remained in the Naqab/ Negev desert in a concentrated region called the Syag (Fig. 2.2) under a military governor. This movement created change in the H'ema; yet even for those who were moved, the territorial affiliation and attachment to this newer territory remain strong. Communities remain as 'closed' tribal areas at the level of towns, kinship or clan affiliations at the level of villages, and extended families at the level of neighborhoods.
• The Bedouin live as a minority within a minority within a minority: They are a minority in the peripheral and marginal Negev region, as well as being a minority among the Arabs of Israel, who are a minority in the Jewish Israeli State. This increases the societal and geographical marginalization of the Negev Bedouin and contributes to strengthening the traditional system of tribalism as a way of protecting their individuality. The situation and status of minorities have a direct impact on the Bedouin as a collective, subcollective-tribal entities, and as individuals. This impact is evident in their minimal economic opportunities, lack of political power, and only nominal and more recent ability to mobilize.
• There is differentiation and fragmentation (religious, cultural, and regional) among Israel's Arab population which Israel's policies reinforce. The separation of the Bedouin living in the Negev, Galilee, Triangle region, and mixed cities contributes to the added marginalization of the Negev Bedouin, as those living in the remotest region (Fig. 3.1).
• The physical planning of Bedouin new towns dates back to the mid-1960s when the government plan for the new town of Tel Sheva was developed, followed by Rahat at the beginning of 1970s. At that time, the towns' physical layouts were based on tribal affiliation and traditional social stratifications at the level of districts and neighborhoods. These residential plans reinforce the tribal and clan affiliation. The plan rationale was to create a tribal town, divided

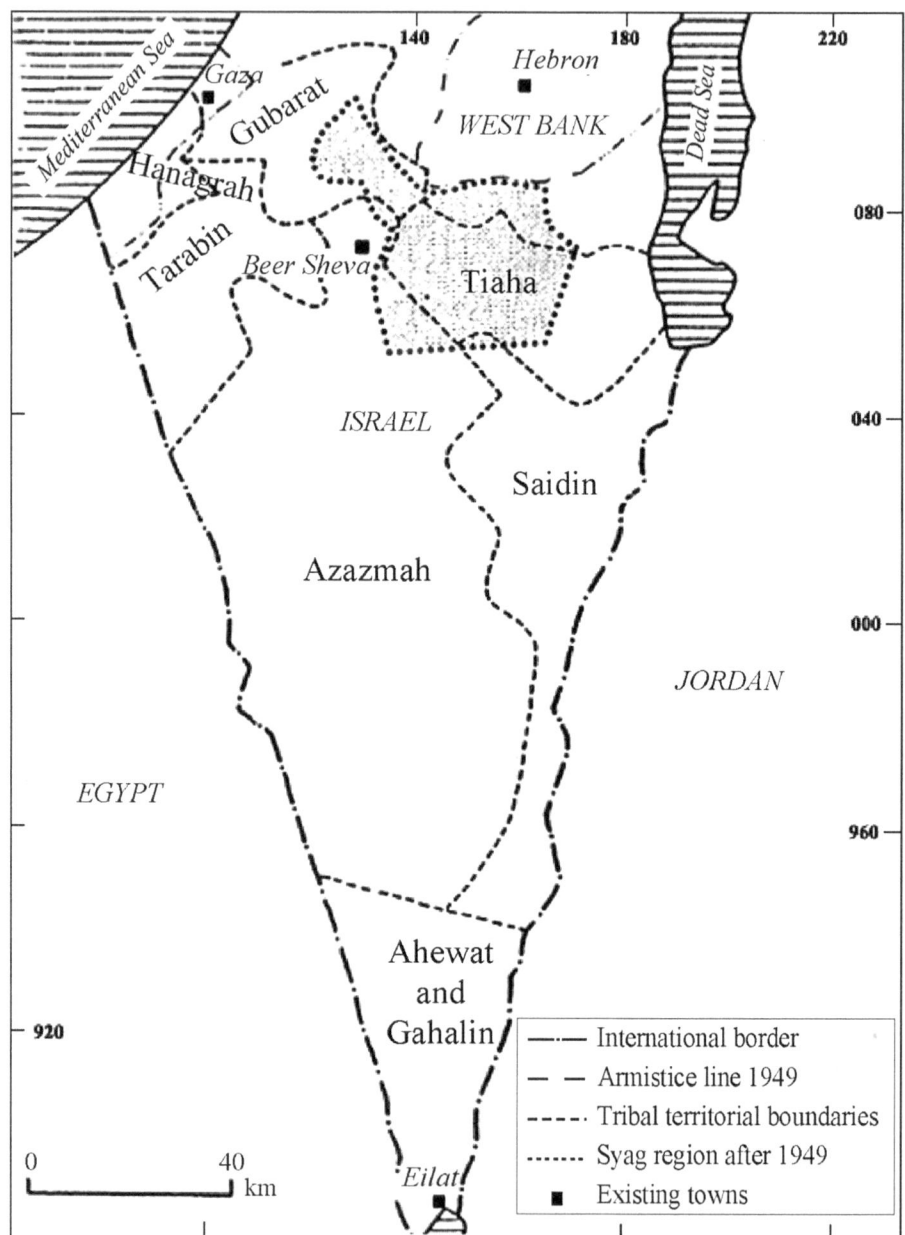

Fig. 2.1 Distribution and territory (H'ema) of the major Bedouin tribes in the Negev 1948, and Syag region (based on Shemony 1949)

Fig. 2.2 Syag region

into clan and extended family neighborhoods, which would allow non-landowners from different clans to congregate. The new planned towns used the Bedouin social structure and hierarchy as a basis for the physical layout (Stern and Gradus 1979).

- The intensity of the ongoing dispute between the State of Israel and the Negev Bedouin over land ownership, resettlement, spatial concentration, and economic development policies reinforces the attachment of tribes/clans to their place and land. This attachment strengthens their sense of, and need for, clan affiliation as a way of protecting both the nuclear family and individuals. The dispute is rooted in the Jewish Zionist and democratic nature of the state, and the Bedouin belonging nationally and culturally to the larger group of Arab inhabitants of the land, who became a weak minority with the establishment of the State of Israel in 1948. These sociocultural and geopolitical differences fuel the conflict, reducing common interests between the state and the Bedouin community (Nasasra et al. 2014).

- The governmental policy of Judaization of the Negev and attraction of Jews to the Southern region is the reverse of its policy of spatial concentration of the Bedouin. For Jews, policies include individual farms in the Negev, Galilee, and the Judean Hills, which are operated and managed by a single farmer, one family, or a small number of settlers, spread over a large area. Israel has 30 individual farms, most of them located in the Negev whose purpose was to both develop tourist sites and maintain national land from illegal occupation (Bedouin).

Community Variation and Subgroups

As a result of the above factors, the Bedouin community comprises a unique subcultural group among the Arabs in Israel. 'Bedouizem' is now acknowledged to be ethnically and culturally differentiated. Beyond the various tribes and clan associations, the Bedouin community is not monolithic in other ways: some are landowners living traditionally on their land, and others do not own land and in this group some are internal refugees.

The Bedouin, a traditional, patriarchal society, continue to cling to their social frameworks, isolating themselves from other segments of society. The community's stratification in terms of social affiliation, class, and ethnicity has a direct impact on municipal organization. Lack of social mobility reinforces territorial enclosure and impedes cooperation and interaction with other groups. Although Bedouin society is no longer nomadic, the economy of those living in villages remains strongly based upon herding with some farming. Therefore these villagers have need for access to land even when concentrated within small municipal settings. Social stratification among the Bedouin population is based upon the following (Marx 1980; Falah 1989; Meir 1997; Porat 2009):

- *Tribal tradition*[2]: Tribes (*kabela*), sub-tribes (*ashera*), clans (*sebit*), and extended families (*hamula*), each branch living in its separate territory in keeping with customary and habitual Bedouin law: the social structure of the Bedouin stratification hierarchy is dynamic, changing according to size (growth) of the tribal units and time period. For example, tribes are divided into sub-tribes (or clans in the case of the Negev). As these clans grow, they themselves become tribes. Often tribes are fragmented socially and territorially during different time periods and ties to the original tribe diminish. Thus, nuclear families become extended families, extended families become sub-tribes/clans, and these become tribes. Social stratification is still strongly connected to previous generations.
- *Social class*: *Bedouin* (nomads), *humran* (farmers/peasants), and *sumran* (servants): Social immobility, combined with restrictive government policies, reinforces this social arrangement with relatively little mixing of classes. However, social class is becoming more dynamic, it too being connected to time, place, habits, size of group, and economic and political power. The social stratification of the seminomadic period of the 1940s differs from that of today which is influenced by residential hierarchies: planned towns versus organic villages (recognized and unrecognized by the government) and degrees of modernization.
- *Land ownership*: Some tribes originated within the area and hold title (dating back to the Ottoman or British Mandatory periods and their land regimes) to the land on which they live.
- *Social stratification stemming from generational and/or professional achievement*: Tensions between educated, young Bedouin and the older, dominant traditional tribal leaders.
- *Internal evictees/internal refugees*: Bedouin evicted in 1948 from their areas of settlement and grazing lands in the South were relocated to the "Syag" area (see Fig. 2.2) on lands belonging to tribes already living there. They now wish to return to their former lands or, alternatively, to secure recognition and receive land rights where they now live.
- *Status and condition of women*: Women constitute approximately half the population, but the dominance of traditional social restrictions renders them vulnerable and they suffer both economically and socially. The reduced demand for traditional occupations such as weaving and crafts creates pressures for greater educational opportunities for women to enable them to better shape their own destinies.

---

[2] Tribal structure and terms vary from country to country. We have adapted nomenclature to fit the Bedouin in the Naqab/Negev. The sub-tribes and clans in the Negev usually refer to the same classification; clan will be used in this book.

This stratification makes it difficult to achieve consensus within the Bedouin community, as the Bedouin are diverse and their status, economic, and political power structures are in flux. The conflict surrounding land claims, recognition, planning, and appropriate municipal frameworks for Israel's Negev Bedouin, especially those living in dispersed, unrecognized villages, is even more complex. The explicit and implicit policies related to the Bedouin in the Negev contribute to the strengthening of their own sense of ethnic separateness. This separateness, rooted in their historical nomadic way of life which sets them apart from the rest of the Arab minority, has evolved into cultural and ethnic behaviors which partially explain the unique problems of spatial planning, management, and organization.

Today the Bedouin have a complex socioeconomic and municipal structure. As a group they are changing from a marginal minority within Israel as a whole and among the Arab minority to a group involved in both national and Arab minority politics (Abu-Rabia 2012). They now identify more strongly with the larger Arab minority who in turn relate to the Bedouin less as a marginal and peripheral group than as a group whose struggles relating to land policy, planning, and municipal structure are both important and central to the larger Arab agenda and can be integrated into the more general lobbying for policy equity.

Thus, it is important to understand the dispute within the wider context of the Arab minority in Israel—the focus of the next chapter.

Chapter 3
Arab Communities of Israel and Their Urbanization

The Bedouin are a subset of the Arab minority. As such it is important to understand the overlying context of Arab Palestinian (henceforth Arab) communities in Israel, which have changed dramatically in recent decades. Their structure has shifted from the small traditional villages to a more modern hybrid rural-urban type or "urbanized village" in a process that is both general and unique to their environmental circumstances (Khamaisi 2012). This urbanization process has led to incremental changes in physical, socioeconomic, and sociocultural environments within the Arab communities, particularly housing characteristics.

For the most part, Arab Israelis live in separate communities, differentiated spatially and ethno-culturally from the Jewish majority (Al-Haj 1995). After the establishment of Israel in 1948 and their becoming a minority, the Arabs resided mostly in small rural localities in the periphery of the new state. A point of reference for Israeli Arabs is 'an-Nakbah'—'the Catastrophe.' Seven hundred and twenty-five thousand Arabs fled or were expelled from Israel[1] in the fighting that preceded the War of Independence, as well as during and after the war. Many expected to return with the success of invading Arab armies. The Israeli victory resulted in appropriation, not only of private Arab-owned land, but also public lands of the Ottoman and British regimes that had been used by Arabs for farming and grazing. Many abandoned villages were razed to be replaced by Jewish settlements, and access to the remaining Arab villages was limited by land expropriation.

During and after the 1948 war, many Arabs living in urban centers such as Jaffa, Haifa, and Beer Sheva fled or were expelled from the country. The remaining Arab Israeli minority amounted to approximately 158,000 in 1949, constituting about 15 % of the entire Israeli population. With a natural annual growth rate

---

[1] The area upon which the Israeli State was established. Prior to the 1949 Armistice, the British called the land Palestine/Eretz Yisrael, and the Arabs called it Palestine.

© Springer International Publishing Switzerland 2015
D.F. Shmueli, R. Khamaisi, *Israel's Invisible Negev Bedouin*,
SpringerBriefs in Geography, DOI 10.1007/978-3-319-16820-3_3

of 3 %, the Arab citizens of Israel today total 1.7 million or approximately 21 %of the country's total population (CBS 2014, page 89).[2] The ethno-national group of Arabs comprises three religious groups: Muslims, 80.0 %; Christians, 10.6 %; and Druze, 9.4 %. The median age of the Arab population in 2013 was 21.5 years, compared to 31.6 years for the Jewish population. However within the Arab community there are differences: the median age among the Muslim community in 2013 was 21.0 years, for the Christian 34.0 (close to the overall Jewish median), Druze 26.3 years (ibid, pages 100–103) and for the Bedouin (Muslim) community 16.1 years (ibid, page 141).

Aside from the religious-cultural diversity, the Arab population is distributed within four main peripheral and marginal areas: the Galilee region in the north with 53.0 % of the population, the Triangle in the central region of Israel with 23.0 % of the population, the Negev region in the south (primarily Bedouin) with 15.8 %, and 8.2 % in mixed Jewish-Arab cities (ibid, calculated based on data appearing on page 133) (Fig. 3.1).

According to data from the 2014 census (relating to 2013), the Arab populace lives in 135 recognized towns and villages. About 41.9 % of them live in cities or towns (compared to 80.8 % of the Jewish population);[3] 47.7 % live in villages with local councils (compared to 9.7 % of the Jewish population); and 6.4 % of the Arab citizens live in small villages with regional councils (compared to 9.5 % of the Jewish population). The remainder of the Arab population (4.0 %) live in unrecognized villages in the Negev (calculations based on CBS 2014, pages 122–123). This geographic pattern evolved following the war of 1948 and persists today because of the limited mobility among the Arabs who live primarily in villages and rural cultural communities. The Arab population of Israel is undergoing an urbanization process, which is accompanied by population growth, an increase in education (the median years spent at school in 1961 was approximately 1.5 years; this figure rose in 2013 to approximately 13 years), and changes in the economic structure from agricultural to nonagricultural activities (in 1961, about 60 % of the workforce was involved in agricultural production; the rate dropped to 4 % in 2013). The number of commuters to workplaces in Jewish localities (where most job opportunities are located) has increased substantially (Khattaab and Miaari 2013).

The urbanization process within the Arab communities, however, has been minimal, tending to remain rural and traditional (Champion and Hugo 2004; Khamaisi 2004; Tarrabeih et al. 2012). During the period between 1948 and 1966, Arabs were under military rule. This regime reinforced the status of a separate Arab minority and limited spatial and functional mobility. Thus, most of the growth was absorbed within the villages, increasing densities to more urban standards without the accompanying urban functions (urbanism). The leadership role of the traditional Arab institutions was retained.

---

[2] This figure includes the Palestinians with permanent resident status (not citizenship) in East Jerusalem (about 300,000 residents). The CBS has included these residents in their census figures since the annexation of East Jerusalem to Israel in 1967, despite their special status.

[3] This includes the population living in cities of 20,000 and more including mixed cities.

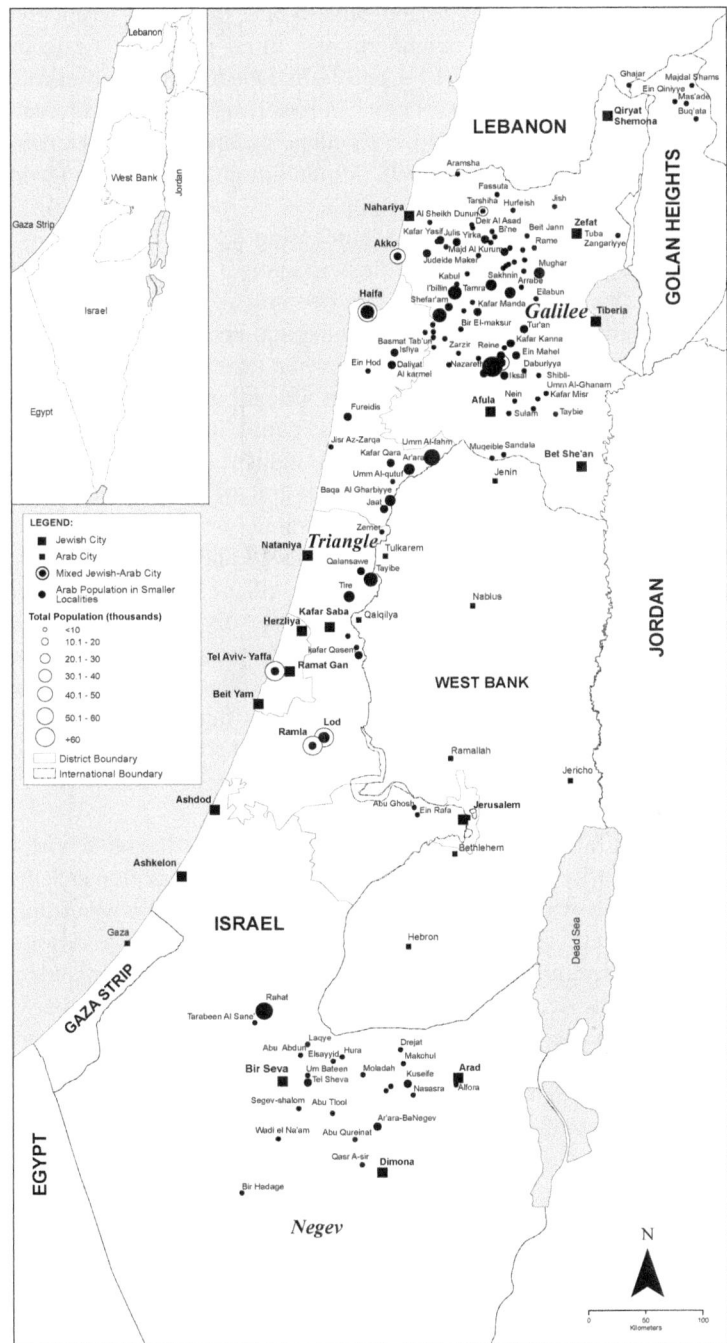

Fig. 3.1 Distribution of the Arab population in Israel, 2013 (not including the unrecognized Bedouin villages in the Negev)

Post-1967, these patterns continued with Arabs primarily living in separate urban-village localities, and in neighborhoods, often primarily Arab, located in mixed cities/towns, such as Haifa, Lod, and Jaffa. Due to both internal and external political forces, majority-minority relations in Israel, and cultural preferences, Arab Israelis have limited rates of migration to cities, and instead import the city into their communities (Hlihel 2011, 63–80). Population growth in Arab towns, resulting from high natural birth rates, is accompanied by increasing housing demands, limited housing market size, and poor purchasing power. Some towns undergo urbanization without legally changing their municipal status to that of a city. In these cases, the town develops a form of 'in situ urbanization' that is characterized primarily by the entry of urban functions into rural space (Khamaisi 2004; Meir-Brodnitz 1986; Kipnis 1976). In other cases, when towns grow sufficiently to qualify for city status, they develop an 'urban-village' pattern that is familiar in developing countries. This pattern gives rise to contradictory land uses, traditional and rural lifestyles, and cultural, commercial, industrial, and residential patterns. When different forms of land use develop in proximity to one another, tensions arise among neighboring land owners and often the municipal structure lacks the capacity to adequately address them. These competing uses and needs in a dense physical setting constitute the in situ urbanization or urban-village mosaics, inevitably giving rise to conflicts between various segments of its diverse regional population. Consequently Arab towns require additional land for residential development, public buildings of institutions, commercial developments, environmental infrastructure, and public parks. This land is not available in sufficient quantities to allow for planned expansion on the one hand, while on the other some of the Arab communities are unwilling to follow zoning regulations (Yiftachel 1992; Ozacky-Lazar and Ghanem 2003, Khamaisi 2010).

The geopolitical and the national conflict between the Israeli Jewish majority and the Arab minority revolves partially around Arab traditional sociocultural norms. This context, which creates a dialectic relationship between the external matrix of government control and the internal sociocultural traditional norms shared by the Arab minority, has a direct impact on housing demands and land attachment among the Arabs. The reality of conflict and contradiction between state ideology and hegemonic policies and the sentiments of the traditional local native Arab community severely curtail Arabs' sense of belonging within the state system (Kretzmer 2002).

The Impact of Limited Migration on Rural-Urbanization Patterns

Israeli Arabs have usually been faced with limited housing mobility and constraints on territorial expansion due to three factors: limited migration practices, self-housing methods, and their private land ownership. The situation created by limited migration among Arabs means that most people are born and die in the same locality and, in some cases, in the same residential home, often enlarged to accommodate family growth. This situation results in latent urbanization in the Arab localities, which occurs as a result of high population and some economic growth without equivalent growth in municipal functions (Meir-Brodnitz 1986; Kipnis 1976; Khamaisi 2005). Some of the new housing stock is built without building permits, and much of the residential areas suffer from lack of municipal services and weak infrastructure.

The main internal sociocultural factor relating to the strong sense of belonging and attachment to family and kinship is known as the '*hamula*' in traditional patriarchal Arab communities (El-Taji 2007). The hamula is not only a framework of biological kinship relations. It functions as a political and socioeconomic unit vis-à-vis other hamulas and other communities as well. Under current circumstances in Israel, Arab identification with the central government is extremely low (El-Taji 2008), the development of civil society remains limited, and Arabs have been effectively excluded from participating, shaping, and producing public space at the state and regional levels. Instead, hamulas provide people with an important sense of safety and belonging. Arabs have therefore preferred to continue living within the traditional social structure of their localities instead of migrating to urban centers separated from their families and hamula. Arab residential patterns are also characteristic of traditional rural societies, with the demand for housing focused primarily on family compounds within specific localities, and in neighborhoods based on hamula and religious affiliation. The customs within the Arab family enhance parental expectation that children live close by but commit parents to provide housing and land required to support their children's needs. The children benefit from this family commitment and support through which separate homes for them are secured. Migration away from the locality involves social, cultural, and financial costs which the Arabs in Israel still cannot afford. Furthermore, the hamula support structure provides assistance in construction of houses within the villages. For those of the younger generation who pursue higher education in universities and colleges in the cities, the fear of losing this social benefit brings them back to the villages upon completing their studies (Masry-Herzalla et al. 2011).

Moreover, the reasons for the limited migration of Arabs within or to other Arab localities relate to the similarity in development and standards of living among these communities (it would not be a step-up) and the social isolationist traditions of villages. One exception to this phenomenon is people suffering from inter-hamula conflicts and rivalries in their own villages who escape to other villages in search of refuge. There these newcomers have a lower social standing compared to the host village families. In this sense, migration between villages is not associated with an improvement in the standard of living, but rather with suffering.

The migration of small numbers of Arabs to mixed Jewish-Arab cities, though limited, entails their settling down in concentrated, segregated, and often lower class neighborhoods (Khamaisi 2008; Hlihel 2011; Masry-Herzalla et al. 2011). Family compounds in these localities are often clustered and dense, allowing internal intimacy on the one hand and barring external intrusion on the other hand. Street and alley layouts aim to achieve these cultural goals at the neighborhood level.

Public areas are created in a way that renders them part of the domain of a particular clan or religious group. Small clusters of commerce and services, and particularly places of religious worship, are scattered throughout residential areas in a manner that accommodates the clan's needs and in a religion-based residential form. Poor living conditions have discouraged rural Arabs from moving to towns (Khamaisi 2008).

For most Arabs, migration to Jewish towns is unthinkable. This is just as well as some Jewish residents make concerted efforts to prevent Arabs from moving in. In many cases both Arabs and Jews in Israel are alienated from one another and continue to live in segregated localities. The mutual sense of fear and alienation implies that although some Arabs and Jews work together, they still prefer to reside in separate neighborhoods. This is reinforced by governmental discriminatory policies which tolerate the behavior of some Jewish communities that refuse to have Arabs as their neighbors, and Arab hamula behavior preventing non-hamula members from moving into their neighborhood. This contributes to segregation between Arabs and Jews, and has a direct impact on urbanization processes.

Over the last decade, a limited number of Arab Muslim families have started to seek housing in Jewish neighborhoods and localities such as Upper Nazareth, Carmiel, and Beer Sheva, a move that in most cases has met with resistance from many Jewish residents. Such moves have also proven difficult due to the lack of suitable cultural community services in these new environments. However, overall, Arab citizens' strong attachment to community, land, and home has reinforced their tendency toward segregation, with some integration within the realm of economic activity and the provision of services.

Thus, despite the rising standard of living and ongoing urbanization, out-migration among Arabs in Israel remains limited, preserving the local and regional geographic distribution of Arabs. This in turn strengthens peripheral, alongside traditional, Arab culture. The minor increase in Arab migration of the past decade altered neither the behavior nor the discourse of Arabs in Israel. Emerging and existing social trends, such as strengthening localism, preserving kinship territory (H'ema), securing ethnonational and religious homogeneity, and preventing outsiders from entering the place to which they are attached, all explain the limited migration of the Arab populace. Today, the limited and involuntary nature of migration among Arabs in Israel is part of a culture of home and place attachment, to which the Jewish Arab political conflict contributed considerably. In this respect, culturally sensitive housing and community building has not been a focus of Israeli planners.

Within this larger context, the limited migration among the Bedouin is also connected to attachment to place and tribes. In the 1950s through the mid-1960s there was a limited out-migration from the Negev to mixed cities such as Ramle and Lod in the center of Israel and to some Arab localities such as Kfar Kasem and Taiba. These Bedouin emigrants lived in separate neighborhoods at the peripheries of the Arab localities. From the end of the 1960s movement of Bedouin as collective clans, extended families, or nuclear families occurred within the Syag region, and mainly from unrecognized villages and hamlets to new planned Bedouin towns. Since the mid-1990s, this type of emigration has decreased.

Table 3.1 Growth of the Bedouin population in the Southern District compared to the total Arab population and to the total population of Israel (in thousands), 1948–2013

Year	1948	1961	1972	1983	1995	2008	2012	2013
Total population of Israel	872.7	2,179.5	3,147.8	4,037.4	5,612.3	7,412.2	7,984.5	8,134.5
Arab population	156.0	247.2	461.0	687.6	1,004.4	1,498.6	1,647.2	1,683.2
Percentage of Arab population from total	17.9	11.3	14.6	17.0	17.9	20,2	20.6	20.7
Percentage Bedouin from total population	1.5	0.8	0.96	1.1	1.5	2.5	2.7	2.8
Percentage Bedouin from the total Arab population	8.33	7.52	6.59	6.54	8.64	12.3	13.1	13.3
Bedouin population of the Southern District	13.0	18.6	30.4	45.0	86.8	185.5	216.2	224.2

Source: Central Bureau of Statistics, 2014, Statistical Abstract of Israel No. 65, Central Bureau of Statistics, Jerusalem, Table 2.15, page 126 and Table 2.16, page 133

[a]Based on census data after the establishment of the State of Israel; Bedouin in Negev moved toward the West Bank and the policy of concentrating them in Syag area began

[b]Since 1967, Palestinians in East Jerusalem (then about 68,000) were added to statistics related to Arabs in Israel. Most of them still have special status as permanent residents but not Israel citizens

Despite the special context and peripheral situation and status of the Bedouin in the Negev, their relationships and connections to other Arabs in Israel began to strengthen at the end of the 1990s and a representative of the Bedouin in the Negev joined other Arab national and municipal informal institutions. Parallel to this strengthened relationship, the Arab political parties and non-governmental organizations (NGOs) such as Adalla, the Jewish Arab Center for Economic Development, Mosawa, and the Al-Aqsa association became more interested and involved in what was happening to the Bedouin communities the Negev.

The relative percentage of the Bedouin in Israel as a whole, and among the Arab minority, has increased over the years as a result of continuing high birth rates and the decreasing of out-migration from the Negev. The practice of polygamy among the Negev Bedouin exists but is decreasing particularly within the Bedouin towns. The decrease of the relative population percentage among the Bedouin between 1948 and 1961 results from exile of Bedouin to West Bank, and migration to some towns in central Israel during this period. External factors affecting the proportion of Bedouin to the total population is Jewish immigration to Israel, and occupation of East Jerusalem and it annexation to Israel after 1967 (Table 3.1).

Bedouin life is rapidly taking on aspects of modernity (Abu-Rabia-Queder 2006). The majority of the Negev Bedouin live in 7 'urbanized' towns and 11 recognized rural villages. Even those Bedouin living in the unrecognized villages do so in compact built areas, developed organically according to customary laws, clan, and land affiliations. In the next chapter we explore possible theoretical frameworks which may shed light on these transformations.

Chapter 4
Theoretical Context: Justice, Urbanism, and Indigenous Peoples

This chapter aims to frame the theoretical context guiding this volume in a way which promotes understanding, also allowing for comparisons between the Bedouin and other minority communities.

Land Claims of Indigenous People

Historically the Western ideological perspective embedded in planning processes has tended to marginalize indigenous perspectives in planning and decision making, and ignore the role of indigenous participation in planning and political processes. Lane (2002) points out that the rational, comprehensive paradigm in planning has regarded the cultural values of indigenous people as irrational relics. Yiftachel (2001) argues that state planning is not necessarily a benevolent practice, and can be potentially repressive, causing minorities to lose faith in state-building projects.

In recent years, the discourse on planning in general and for indigenous communities in particular (Berke et al. 2002; Hibbard et al. 2008; Lane and Hibbard 2005; Meir 2003, 2005; Sandercock 1998; Zaferatos 1998) reflects a shift from state-imposed prescriptions toward a more participatory approach. Planning practice now includes the need for collaboration among diverse actors (Healey 1997; Innes 1995, 1996; Innes and Booher 1999; Susskind and Cruikshank 1987; Susskind and Field 1996; Susskind et al. 1999). Lane (2005) also contends that planning must include state transformation, as well as indigenous mobilization, because radical planning (Lane 2001) working outside the state machinery is incomplete. Zaferatos (2004) demonstrates this conclusion in his examination of the Swinomish Indian Tribes' realization that it could not successfully achieve its community development goals in isolation from the surrounding political region, and shows how they formed cooperative agreements which advanced tribal interests and helped to achieve Washington State's growth management goals. Dale (1999) presents an anthropological perspective in the planning issues facing the Haida Gwaii in British Columbia

© Springer International Publishing Switzerland 2015
D.F. Shmueli, R. Khamaisi, *Israel's Invisible Negev Bedouin*,
SpringerBriefs in Geography, DOI 10.1007/978-3-319-16820-3_4

and as Merry (Dale 1999, 933) comments, the robustness of this perspective shows just how much framing and context matter.

The declaration on the rights of indigenous peoples was adopted by the UN general assembly on September 13, 2007. There is no universally accepted definition of 'Indigenous,' though there are characteristics that tend to be common among indigenous peoples (Howard 2003):

- They tend to have small population numbers relative to the dominant culture of their country. However, in Bolivia and Guatemala indigenous people make up more than half the population.
- They usually have (or had) their own language. Today, indigenous people speak some 4,000 languages.
- They have distinctive cultural traditions which are still practiced.
- They have (or had) their own land and territory, to which they are tied in myriad ways.
- They self-identify as indigenous.

Some of the above characteristics fit some of the Negev Bedouin, who make an effort to preserve components of traditional Bedouin culture, alongside growing and instrumental modernism. Amara et al. (2012) entitled their article *Indigenous (in) Justice* and focused on Negev Bedouin and human rights laws (Braverman et al. 2014). In addition to the desire of Bedouin to preserve and secure their cultural and social fabric as indigenous people, they desire to participate in general regional and national policy and development, and often meet barriers to doing so.

In their study, *Addressing the Land Claims of Indigenous Peoples*, Susskind and Auguelovs (2008) offer a set of findings and principles designed to help indigenous people protect and preserve their cultural and territorial rights. The authors' thesis is that native people must be allowed to pursue their lives in a manner consistent with their culture, traditions, and lifestyles, recognizing their unique relationship to the land. They argue that taking away the lands of an indigenous people or forcing urbanization upon them is equivalent to their destruction.

Among the principles, derived from 14 case studies of indigenous people are the following:

- Land is fundamental to survival. Indigenous people must have a place of their own where they can live together, separated from the dominant society by geopolitical, symbolic, and cultural boundaries.
- They must have some degree of autonomy, especially over management of their resources and administration of public services. This involves preventing economic patterns that endanger their way of life, and the right to decide on their own economic activities.
- United internal leadership and cohesive voice are necessary to preserve or regain cultural and territorial rights, and win the support of national and transnational civil society networks. This requires the building of strong internal organizational capacities through the creation of representative bodies and improved leadership. If negotiations fail or are inconclusive, the need to turn to litigation requires substantial outside help.

In several of the cases cited, indigenous peoples have succeeded in having direct negotiations with national and local governments and resolving their land claims. These include the Nisga'a of British Columbia, the Bedouin of Jordan, the Pima of Arizona, and the Nama of South Africa. The population of these groups is very small (11,000 and under) with the exception of Jordan, where the estimates of current Bedouin range from 50,000 to over 100,000.

In some of these cases, indigenous people maintain little or no contact with the outside world. This is not so with either the Jordanian or the Israeli Bedouin, both undergoing urbanization processes, or exposed to them. In Israel their lives have evolved in reciprocal relationship to Israeli urban centers. The Black Goat Law of 1950, designed to prevent land erosion, made most grazing on Negev lands illegal, accelerating the shift toward farming, working in construction and other urban pursuits, and volunteering for the army (Marx 2005). The Jordanian Government has taken a collaborative approach in its negotiations with the ten Bedouin tribal groupings that are mainly located in the south (Wadi Mujib), who wish to maintain their traditional way of life. The constructive dialogue between them and government has resulted in their achieving certain goals such as registration of Bedouin land claims, agricultural and technical assistance and loans, wells, and the set-aside seats in the Jordanian parliament. This is made possible by the Bedouin origins of most Jordanians, although they long discarded nomadism. This excluded the nearly two million Palestinian refugees who represent over 1/3 of the Kingdom's total population. Of equal importance is the political reality that, while Bedouin no longer are the dominant force within the 70,000-man Jordanian army, they continue to play a leading role. In addition, they have their own police force in their tribal areas. In fact, the Jordan Arab Legion, the Bedouin military force, first organized by the British Glubb Pasha, was the backbone of the Hashemite Royal family, playing the key role in Jordan's invasion of Palestine in 1948, and in putting down the Palestinian attempt to overthrow the King in the "Black September" uprising of 1970 (Shmueli and Khamaisi 2011).

In Israel the relationship between the Negev Bedouin and government is very different, sharing neither common background nor strategic interests. Maintaining Bedouin traditional lifestyle within a highly developed urban country is a challenge for both. Certainly some of the principles set forth in the Susskind and Anguelovski study are applicable to the Israeli case. But challenges to Bedouin unity are not only divisions among and within different tribes and their clans, but they also relate to the fact that most of today's Negev Bedouin are descended from three different main groupings—Tarabin, Tiaha, and Azazmah—indigenous Arab nomads, fellaheen from cultivated areas who became nomads, and those who were brought from Africa as slaves. While most Bedouin came to Palestine from Sinai, some migrated from lands east of the Jordan River, and others trace their ancestry back to Yemen. Prior to modern national state borders, the Negev, Sinai, and Jordan were one large open space and Bedouin roamed from one area to another according to customary laws and habits.

Resettlement of indigenous peoples has occurred in different parts of the world. Some have been voluntarily resettled when common interests have existed between central governments and the communities. In other cases imposed resettlement—using legal, economic, and institutional tools as well as state power—has occurred, often with resistance and counteractions from the minority. An example of forced resettlement of Negev Bedouin is the village of Al-Arakib. Al-Arakib is located seven kilometers north of Beer Sheva. It is an unrecognized village and as such does not appear on official maps, nor is there any official sign marking its existence. This village has been demolished and rebuilt more than 70 times. The government does not want to recognize the village, and did not accept the land claim of the Ul-Oqbe clan. The clan went to court and in 2012 the court denied the land claim which was required if the Ul-Oqbe clan was to plan and develop a new recognized village on Al-Arakib. The Jewish National Fund planned to include the land of Al-Arakib in a new forestation project in the Negev desert called *Ambassador's Forest*, honoring the assistance provided to Israel by the world's diplomatic corps. This plan spurred objections from Bedouin living in the town of Rahat and several villages nearby. During the court appearances, a large number of stakeholders were involved including advocates from academia for both sides. This is an example of the problems of resettlement faced by the remaining 36[1] unrecognized villages in the Negev.

The issue for Israel's Bedouin, as it is for many indigenous peoples, is not political sovereignty as an independent state, but the type of territorially framed autonomy that is realistic. This devolves on what kind of autonomy Israel is prepared to grant, given its concerns about national security; land pressures from nearby Israeli cities, towns, and settlements; infrastructure costs; and economies of scale. This clash is expressed in a variety of ways, but especially over land boundaries, issues of ownership or catchment (historical usage), and control of land-use planning.

Urbanism, Modernization, Westernization, and Forced Urbanization

The world is rapidly urbanizing. This accelerated trend (Pugh 1995; Abu-Lughod 1996) raises a number of questions regarding how populations cope with urbanization processes. Despite urbanization being a global phenomenon, most of the theories relating to this process are derived from Western (North American and European) thought and connected to the concepts of modernization and Westernization. These explicit and implicit connections generally lead to the imposition of a Western model of urbanization on communities with a limited cultural or contextual Western orientation such as the Negev Bedouin in Israel.

---

[1] The number of unrecognized villages differs according to source. The figure around which there is the greatest consensus is that 11 villages have been recognized by the government and managed by two regional councils leaving 36 villages awaiting recognition. Other sources claim greater or fewer remaining unrecognized villages according to interests, definitions, and clans.

Modern urbanization has had significant impact on the physical and cultural transformation of traditional rural settlements (Berry 1976; Mai and Shamsuddin 2008), as a result of transforming villages to towns, or expanding urban settlements to include neighboring villages as part of urban sprawl (Bourne 2001).

As cities develop, agricultural land is reduced; what were once primary are transformed into secondary and tertiary occupations; population densities and costs of land increase; high-rise construction technologies are adopted; and transportation infrastructures are developed within and between urban centers, metropolises, and peripheries, to accommodate commuting from the suburbs. Furthermore, cities require service systems—e.g., sanitation, healthcare, education, and welfare—which, in societies that have not developed measures to cope with these intense changes, are ill prepared to manage population growth.

The two main leading forces in urbanization relate to population growth, the first being natural population growth and the second immigration. These forces operate within a framework of economic change which affects patterns in community life and the rate of investments in public and private services. They are accompanied by changes in employment and housing demands (Khamaisi 1996, 2009; Kafkoula 2009). The rapidity of population growth places a strain on a community's ability to provide adequate housing. The increase in residential density affects land allocation policies as well as greater demand for public services, commercial activities, and job opportunities.

There seems to be a misguided yet widespread conception among researchers, planners, and policy makers that there is a direct correlation between modernity and urbanization, and Westernization. This assumption has led to frequent attempts to foist Western models on other societies without taking into account their specific cultural and spatial contexts. Such models are the products of shared industrialization and urbanization processes as well as modern patterns of behavior. Traditional residential village patterns were affected by such factors as construction technology, availability of building materials, community customs and traditions, aesthetic values of the society, and small-scale production (Cannana 1933; Akbar 1988; Al Hathlol 1994; Elshesawy 2008). The changes often emerge from the need to provide mass housing production and sales to meet the needs of population growth. This may transform villages into towns and expanded cities. Today more than half of the world's population lives in urban centers (United Nations 2008), and currently among the Palestinians in the West Bank and Gaza Strip, over 65 % live in urban areas (Abu Sada and Thawaba 2011). High immigration rates and population concentration in such areas lead to a high demand for housing, which in many cases creates shortages (Song et al. 2008).

The urbanization of villages, changes in community economic structure and growth, building technology, and building styles and architecture as part of globalization, all contribute to reducing the traditional dichotomy between rural and urban areas (Champion and Hugo 2004; Henderson and Wang 2005; Khamaisi 2012). Numerous villages worldwide have become urbanized; they grow demographically and physically, but still operate according to traditional patterns. This structure has been referred to as an urban village (Wang et al. 2009). The emergence of the urban village style is part of the urbanization process. Meeting housing demands through

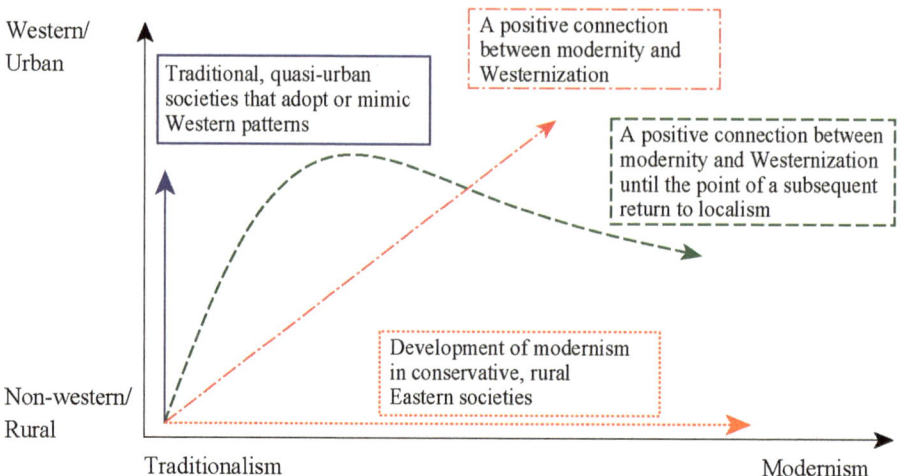

Fig. 4.1 Range of settlement forms on the modernism/traditionalism and urban/rural axes. *Source*: Khamaisi 2012

mass production contributes to developing the phenomenon of the urban village as an integral component of urbanization. Limited migration and attachment to place (Hidalgo, and Hernandez 2001) contribute to developing urban/rural communities (Eben Saleh 2002) where only a limited degree of urbanism occurs. By urbanism we refer to functions such as infrastructure efficiency, transit-oriented development, institutional integrity, regional integration, and economic and cultural advantages to economies of scale which are assumed to accompany urbanization, but in the case of the Arab population in Israel, including the Bedouin, these functions are lacking.

However, there are other settlement types that emerge from non-occidental experiences. As Huntington (1996) opines in *The Clash of Civilizations* one such model is informed by the use of modern Western goods and technologies along with a propensity for traditionalism and conservatism. In our estimation there is room for other models.

Figure 4.1 presents communal patterns of behavior and links between urban/rural and modern/traditional. Some conservative rural societies continue to dwell in villages, but transform themselves into a modern populace by consuming contemporary products and technologies. Alternatively, other traditional rural societies undergo urbanization and imitate urban Western patterns of behavior. In both these societies, tensions surface between traditionalism and urbanism. Some urbanizing communities embrace modern norms, but the transformation reaches the point where the populace fears losing customs and values which it wishes to safeguard and perpetuate. As a result, the residents cease to mimic the paradigms of Western urbanism. Despite continuing to embrace up-to-date goods and technologies, the inhabitants maintain conservative local and cultural behavioral patterns. In light of the above, as planners we realize that urbanization does not necessarily point to a shift toward modernity and Westernization, certainly not in all realms of life, and that the robust consumption of sophisticated goods is quite capable of coexisting with conservatism and traditionalism.

In the context of an emerging Israeli State forged by *Ashkenazi* (European) Jews, Western settlement patterns were imposed on *Sephardic* Jews (from Arab and Islamic countries) as well as on the Israeli Arab minority including the Bedouin. The goal was to create a modern Western nation. This goal produced policies of both securing greater integration on the one hand, and territorial, demographic, and social engineering on the other.

When Western modernization is imposed from top to bottom it often benefits physical planning of housing and supply of services, but limits considerations of multiculturalism and diversity. In traditional nomadic, seminomadic, and rural graz-ing/farming communities, villages developed organically with few demands for either new housing or functional services. Traditional residential village patterns were affected by such factors as construction technology, availability of building materials, community customs and traditions, aesthetic values of the society, and small-scale production (Cannana 1933; Akbar 1988; Al Hathlol 1994; Elshastawy 2004; Elshesawy 2008).

Urbanization is at times a spontaneous process or phenomenon, and often a com-ponent of state planning policy to reorganize the settlement map and direct popula-tion distribution. When a component of state planning, it sometimes involves forced urbanization. This process seeks to protect agricultural lands, reduce or limit rural and urban sprawl, shift demographic trends, achieve economic and functional devel-opment, and secure environmental resources. Frequently when imposed, urbaniza-tion occurs without urbanism. This means that the physical fabric of the locality changes but is not accompanied by personal and community sociocultural and behavioral benefits.

In Israel, forced urbanization is a consequence of governmental, territorial, and planning policies and has major implications on resettlement and spatial concentra-tion of the Bedouin in the Negev.

Justice: Distribution, Recognition, Participation, Compensation

For Arab citizens of Israel, environmental justice is inextricably tied to the more general demand for treatment equal to that accorded Jewish Israeli citizens, and restoration of lands lost in the wake of Israel's War of Independence and later expro-priations. For the Bedouin, it is particularly acute (Amara et al. 2012).

Most established theories of environmental justice focus on environmental aspects of distributive justice (outcomes); some focus on processes which lead up to and affect justice issues and decisions. We will adopt a holistic approach developed by Shmueli (2008) which encompasses distributive issues as they relate to land allocation and use within the unique Israeli context, adding as process elements participation, recognition, and compensation.

Israel has a special situation that is distinct from general macro-environmental risks and benefit inequalities. This is a category of conflicts which arise from limitations to Israeli Arab political and civic equality within a state that is defined as a 'Jewish State' that is politically, economically, and socioculturally structured to ensure the dominance of the Jewish majority. As shown in Chap. 3, the majority of Israel's Arab population (1.7 million) is concentrated in the northern (Galilee) and southern (Negev) peripheries, Wadi Ara's Um el Fahm and the rest of the 'Little Triangle' outside of the coastal plain as Israeli citizens, and East Jerusalem (as permanent residents). It is in these peripheral regions that issues of environmental (in this context particularly as relating to land) inequities for Arabs are most sharply felt.

The discourse over land allocation and use that takes place between Israeli Arabs and Jews is heavily influenced by the fact that the two peoples frame the equity issues from different perspectives and narratives. Most Israeli Jews consider the petitions for land expansion by Arab communities within the context of national and local security needs. For the Arabs the need for additional land stems from demographic and socioeconomic pressures made more acute by feelings of injustice owing to historic claims to lands which they once owned or used.

An example of the differing perspectives toward land is the National Master Plans for Nature Reserves and Forests. They are seen by Jews as preservation tools that also provide security benefits. Arabs view these national plans as regulations designed to limit their ability to expand in order for them to maintain the geographical cohesiveness of families and clans, and maintain their cultural distinctiveness (Shmueli 2008). The Negev example of the Ambassador's Forest mentioned earlier is a case in point. Limited opportunities to participate in processes that lead to development of national and regional master plans fuel Arab feelings of discrimination and injustice.

Oren Yiftachel has taken a leading role among Israeli geographers in criticizing Israeli planning policies that affect the Arab populace unfairly (Yiftachel 1999). He faults the National Plan for authorizing 30 Jewish private farms spread over a swath of the northern Negev plateau. He rejects the initial rationalizations for the Plan as a means of preventing the Bedouin from taking over Negev land, and then as a means of stimulating tourism. Advocating the need for planning authorities to adopt proper planning policies, he calls for recognizing the rights of the Bedouin of this region to equality and distributive justice (Yiftachel 2006).

There is no question that there is a differential in both the abilities of the Jewish and Arab communities to cope internally with hazards posed by solid waste disposal, sewerage, toxic waste, noise, foul smell, well water pollution, soil erosion, forest fires, and air pollution from quarries—and with governmental support in their amelioration. Most Jewish Israeli communities are able to levy local taxes which supplement state revenues, while Arab communities depend for the most part on state support for their budgetary needs, which are generally insufficient to address many of the environmental risks. These communities make only limited efforts to levy local taxes, but even when they do so, their tax collections yield far lower revenues. This inequity in economic capacity is a major basis for the Arab claims that the absence of compensatory governmental support is unfair.

An example of a particularly troublesome case for a Bedouin community is the siting of a large chemical complex and hazardous waste facility at Ramat Hovav in the northern Negev, 500 m from the Bedouin encampment of Wadi el Na-am. There are ongoing negotiations regarding resettlement of this unrecognized village due to health dangers.

In determining whether the current recognized Bedouin towns and planning for the recognized (but yet to be planned) or unrecognized villages fall within the definition of environmental (in-) justice, given the Israeli parameters, we will use the fourfold approach (distribution, participation, recognition, and compensation) which makes plausible a pluralistic yet unified theory and practice of justice.

The theories presented will be used to analyze the situation of the Negev Bedouin—indigenous people experiencing rapid urbanization and modernity in gray (Yiftachel 2009) situations, and the impact of government tools such as spatial planning, municipal reorganization, and land ownership on community control, and community responses.

Chapter 5
Negev (in Hebrew) or Naqab (in Arabic) Bedouin

The Bedouin of the Negev: A Short History

Palestine was part of Syria (Bilad al-Asham), ruled by the Ottoman Empire from 1516–1918. Ottoman legislation, especially with regard to land, was based on Islamic Sharia' law. Not until the second half of the nineteenth century did the legislation address individual land rights in the modern sense. Generally, land which fell under Ottoman rule was considered property of the Sultan (state-owned land) and was leased to rulers of provinces and taxes were exacted on crops grown.

On April 21, 1858, the Turks enacted the Ottoman Land Law requiring that the names of landowners be officially recorded as a means of regulating land-related matters in the Ottoman Empire. According to the law there were five categories of land: *Mulk or Rikbah* (land under private ownership, including buildings and crops on the land), *Miri* (Sultan- or state-owned land that could be cultivated for a one-time fee), *Mauqufa or Waqf* (land in a religious trust or Islamic endowment, subject to Sharia' law), *Metruka* (uncultivated land, such as public roads or pastures belonging to several villages), and *Mawat* (wasteland unsuitable for cultivation, usually at least half a mile away from villages). Most of the land in the Naqab was categorized as Mawat. Land issues were settled through verbal descriptions of land boundaries, not measurements or maps (Ottoman Land Law, article 47). If an individual held Miri or Waqf lands for 10 consecutive years with no objection from others, he was entitled to get a free title to the land (Ottoman Land Law, article 78) (Kedar 2001). Two additional laws were passed in 1858 and 1860 regulating land titles (*Kushan*). When a holder of Miri land desired to transfer it to another party, he was required to obtain a signed certificate from the *Imam* or *Mukhtar* (leader) of the neighborhood or village stating (a) that the holder had documentation of ownership; (b) the amount of money paid; (c) the district and village within which the land was located; and (d) the details regarding size and boundaries. However in general the accuracy

© Springer International Publishing Switzerland 2015
D.F. Shmueli, R. Khamaisi, *Israel's Invisible Negev Bedouin*,
SpringerBriefs in Geography, DOI 10.1007/978-3-319-16820-3_5

of the Kushan was limited both due to the lack of accurate maps and the continuous change in names over the generations. During the Ottoman period only about 5 % of Palestinian territory was registered, including a small area in the Naqab (Hezmawe 1998).

The illiterate Bedouin of the Naqab were opposed to creating a written record of their land holdings, since doing so would make them subjects of foreign rule, requiring them to pay taxes and serve in the Ottoman army. Between the years 1912 and 1913 several temporary land laws were enacted although not approved until the outbreak of the First World War. Between 1918 and 1920 the British military followed the Ottoman Land Regime until the British Mandate commenced in 1920.

The British Land Ordinance of 1920 passed land management from the military to the British civil mandate. It primarily adopted the Ottoman Land Laws, cancelled transfer restrictions or land purchase by association, and paved the way toward 'foreign' land purchasers. In 1921, the British Mandate Government issued an order calling for residents of the Naqab to register their land. The Bedouin, who were given a 2-month extension, did not do so, and their land remained unregistered. According to the Land Ordinance (Mawat) of 1921, a Bedouin who cultivated, revitalized, and improved Mawat land was given a certificate of ownership for that land, which was then re-categorized as Miri. Few agreed to the provisions of the ordinance. In 1926 the Mandate provided another 3-month period for land registration. Still, few registered. In 1928 the Mandate began to apply cadastral methods to land registration—based on precise measurements and maps rather than based on verbal description of boundaries.

Twenty-seven years later the courts of the State of Israel ruled that any Bedouin who passed up the opportunity to register Mawat land in 1921 and did not receive a certificate of ownership was no longer eligible to do so (Ben-David 1996).

While many of the Bedouin had not registered their lands in the Ottoman or British land registries as other Arabs did, both the Ottomans and the British respected the customary rights the Bedouin had to the land and did not attach importance to the arid, desert terrain of the Naqab classifying it as Mawat (dead) land, which was not economically viable.

Until the establishment of the State of Israel in 1948, the Bedouin were, for the most part, the sole residents of the Naqab (in Hebrew Negev) desert. In 1947, over 95,000 Bedouin, members of 96 different clans, lived in the arid expanse stretching southward from Kiryat Gat and Ashdod to the Gaza Strip and Sinai. According to several sources (e.g., Al-Dbag 1988; Falah 1989) these Bedouin held approximately two million dunam of land, for which they adhered to a clear, agreed-upon system of property rights. The land was divided according to intertribal agreements, known as H'ema. This land was an important resource, intricately linked to their culture and way of life. Some Bedouin were seminomadic, using the northern Negev to cultivate crops, and moving around with their herds for grazing. The majority lived farther south as nomads in the heart of the desert, where they were limited to grazing their herds of goats, sheep, and camels, along the same annual circuitous pathways (Shmueli 1980) (Table 5.1).

Table 5.1 Growth of major Bedouin tribes in Negev during the British Mandate period, according to censuses of 1922, 1931, and 1946

Tribe	Numbers		Percentage	
	1931 Census	1946 Census	1931 Census	1946 Census
All tribes	47,981	95,556		
Azazmah	8,661	16,505	18.1	17.3
Hanagrah	3,756	7,281	7.8	7.6
Gubarat	4,432	4,274	9.2	8.7
Saidin	639	890	1.3	0.9
Tarabin	16,330	33,064	34.0	34.6
Tiaha	14,163	27,069	29.6	28.3
Ahewat and Gahalin		2,483		2.6

Sources: Atzmon 2013, who based his figures on the following sources:
1. Aref el`Aref, Al Qadan Bayn al Badw, Jerusalem, 1351/1933
2. Barron, J. B. Palestine, Report and General Abstracts of the Census of 1922, Jerusalem, 1923
3. Bromberger, E, The Growth of the Population of Palestine, *Population Studies*, 2(1), 1948
4. Dajani, S. W, The Enumeration of the Beersheba Bedouin in May 1946, *Population Studies*, 1(3), 1947
5. Government of Palestine, Department of Statistics, A Survey of Social and Economic Conditions in Arab Villages, 1944, *General Monthly Bulletin of Current Statistics*, July 1945
6. Loftus, P. J., Features of the Demography of Palestine, *Population Studies*, 2(1), 1948
7. Mills, E., Census of Palestine 1931, Vol. 1, Alexandria, 1933
8. Muhsam, H. V., Fertility and Reproduction of the Bedouin, *Population Studies*, 4(4), 1951
9. Muhsam, H.V. (1951). Some Notes on Bedouin Marriage Habits, *Proceedings of the 14th International Congress of Sociology*, Vol. 4, 297–316
10. Muhsam, H. V. (1956) Fertility of Polygamous Marriages, Population Studies, A Journal of Demography, 10(1), 10–12

With the 1948 Israeli War of Independence[1] many Bedouin fled or were expelled from their lands and the Bedouin population was reduced to about 13,000 (Ben David 1983; Central Bureau of Statistics (CBS) 1949). In 1951 Bedouin in the Negev numbered approximately 12,740 persons, belonging to 17 remaining clans (Ben David 1983, 55). For the newly created State of Israel, the land of the Negev was seen as an integral part of the country and Jewish immigrants were encouraged to settle there.

An important change involved Mewat land, defined under the Ottoman Land Code as 'dead land,' uncultivated and distant from towns and villages (Kedar and Yiftachel 2006). The Arab claims to such lands were taken into consideration by the Israeli Supreme Court in the 1950s and 1960s, when rules regarding Mewat lands were significantly reinterpreted. Whereas the Ottoman Law entitled people proprietary rights to the Mewat lands that they had cultivated, and British Mandate Law allowed opportunities for official registration of this land, the Israeli application established Mewat land as state property (Kedar 2001, 952–53). Israel did not recognize Bedouin ownership of the land in the absence of land registration documents

---

[1] Referred to as al Nakba—the Catastrophe—by the Arabs.

issued by either the Ottoman or the British. Consequently much of the land, held by Bedouin Arabs for generations, was confiscated by the state, most of it in accordance with the Israeli Land Acquisition Law of 1953.

A large majority of the Bedouin remaining in the Negev after the 1948 war, and who subsequently became citizens of Israel, were relocated to the Syag which covered only 10 % of their former territory (see Fig. 2.2).

Well over half (146,700) of the current Bedouin population of 224,200[2] (CBS 2014, Table 2.13, 119) have been relocated to towns, 15,100 of them are in recognized villages organized within two regional councils, and 6,700 live in the cities of Beer Sheva, Arad, or Yeruham. The remaining 55,700 live in 36 unrecognized villages which are regarded as 'illegal' pursuant to the 1965 Planning and Construction Law, and which requires all construction to be subject to a permit for building. Since they lack the necessary licenses (whether in unrecognized or most of the villages recognized as part of regional councils since 2003), residents live under constant threat of the demolition of their homes, often shacks with tin roofs, which become unbearably hot in the summer and do not provide adequate shelter during the cold winters. Nonrecognition is a means employed by the state to force the residents off their lands and concentrate the Bedouin within the government-designated urban enclaves. The unrecognized villages lack basic infrastructure facilities and services such as electricity, running water, roads, healthcare, and kindergartens. These Bedouin are repeatedly assured by the Government of Israel that they will receive all the facilities and services they require when they move to the government-built townships (Centre on Housing Rights and Evictions (COHRE) 2008).

The Bedouin have a very high annual growth rate. In 2013, the annual growth percentage among the Negev Bedouin was 3.7 %, compared with 1.4 % among the Negev Jewish population, 1.9 % among the total population of Israel, and 2.2 % among Israel's Arab minority (CBS 2014, Table 2.13, 119). The rapid population growth among Bedouin creates challenges for the community itself, for municipal structures, and for the state in providing services—including urbanization and resettlement.

The Context for Israeli Territorial Expansion

Displacement of the Bedouin from their lands must be seen within the context of Israel's policies designed to control these lands and transform their ownership (Kedar and Yiftachel 2006). Yiftachel (2002) uses the phrase "Judaizing the

---

[2] Demographic statistics about the Bedouin population are highly contested and different statistics are used by different groups (even different government publications vary) to promote different agendas. This book uses the Statistical Abstract of Israel, 2014, Central Bureau of Statistics, relating to the 2013 end-of-year data.

homeland," to denote the process of restructuring this territory pursued by the Israeli Government since 1948. Driven by the premise that the land belongs to the Jewish people, based on their presence during the Biblical times, an "ideological and moral project" was instituted to populate the country with a Jewish majority.

Complex institutional and legal mechanisms made possible the confiscation and occupation of land; Jewish-owned lands within the State of Israel increased from the pre-1948 percentage of 6.4 % to holdings of 93 % (as of 2013). Confiscated properties are deemed state lands, falling under the ownership of organizations such as the Jewish National Fund, the Jewish Agency, and the Zionist Federation (Amiad, and Moshkin 2002). The transfer of lands to these bodies represents a "black hole" (Yiftachel 1999), from which Arab-owned lands are nearly impossible to retrieve. As Sandy Kedar observed, the Israeli legal system transformed land possession rules in ways that "undermined the possibilities of Arab landholders to maintain their possession," and enabled the transfer of ownership to the Jewish State (Kedar 2001). Here the modern Western legal system, with its cache of technical and scientific language, legitimizes confiscation of land from the Arabs.

The Israeli spatial policy of Judaizing space is applied to the Negev through a matrix of control which includes three components—land ownership, spatial planning, and land management and spatial regime-dividing by municipal jurisdiction. The policy aims to limit and shrink the Arab areas through land confiscation, regulatory planning, housing demolition, and concentration of Bedouin populations. This policy has created a clear system of segregation, which began with the building and expansion of the Israeli settlements, the confiscation of Bedouin lands, and the creation of dual-municipal systems.

Chapter 6
Evolution of Local Authorities: A Historical Overview

An internal barrier to Negev Bedouin resettlement is the current lack of capacity on the individual village level to govern. This feeble capacity is indicative of weak socioeconomic localities, a category to which most of Israeli Arab and all Negev Bedouin municipalities belong. Here too, the wider context of local government in Israel has to be examined in order to understand the Bedouin challenges.

Local government was long established in the region before the State of Israel was created in 1948. Formally, the modern local government was first introduced to Palestine during the Ottoman rule with the enactment of the Vilayet Law of 1864, which provided for the establishment of *nahiyas* (subdistricts) throughout the country. Nahiyas were gradually introduced between then and World War I. Under the law each nahiya was to have a local council but few were actually established. Instead, mukhtars were installed to replace the sheikhs who headed the local hamulas (Al-Haj and Rosenfeld 1990). Under the law, two mukhtars were to have been elected in each village along with a council of village elders (*ikhtiyariyya*) but in fact most mukhtars were appointed, usually after consultation with the local notables. The mukhtar was responsible for assessing and levying taxes among the villagers, settling local disputes, and acting as intermediaries in the relationship between the provincial administration and the village (Elazar and Kalchheim 1988). In some places, particularly Arab local authorities, competition existed between the traditional kinship systems based on patriarchic hierarchy which included mukhtars and sheikhs and the modern system of local authority based on elections (Khamaisi 2008).

Municipal government was introduced at approximately the same time as rural local government and was also a product of the Ottoman reform (*tanzimat*). Jerusalem was made a municipality by special imperial decree (*firman*) in 1863. It remained the only formally organized municipality until the enactment of the Provincial Municipalities Law in 1877 under which 22 towns and larger villages were granted municipal status in the 1880s and 1890s. According to this law, the Ottoman rulers began to establish municipalities in the largest towns of Palestine. The municipalities had an impressive list of powers and responsibilities but, in fact,

© Springer International Publishing Switzerland 2015
D.F. Shmueli, R. Khamaisi, *Israel's Invisible Negev Bedouin*,
SpringerBriefs in Geography, DOI 10.1007/978-3-319-16820-3_6

had very little room to maneuver under the provincial authorities. Municipal budgets were very small and the municipalities had almost no civil service. Municipal councils (*majlis umumi*) of 6–12 members were to be elected by local taxpayers who were Ottoman subjects. In fact, genuine elections rarely took place. The general rule was that the local notables would agree upon a slate of council members. At the outset of the British Mandate, there were 22 municipalities in Palestine: 16 Arab and 6 mixed Arab-Jewish (Jerusalem, Jaffa, Haifa, Tiberias, Safed, and Hebron). Tel Aviv was still regarded as a suburb of Jaffa; however, it was administered by a separate autonomous Jewish council until 1934 when it was declared as an independent municipality (Elazar and Kalchheim 1988).

The British Mandate more or less accepted the Ottoman system until it could introduce a municipal framework of its own. In the interim, new municipalities were created by the Mandatory rule through orders-in-council. The first independent step by the Mandatory rule was the promulgation of a town planning ordinance in 1921, a reflection of the British interest in the aesthetics of the Holy Land. A local council ordinance passed the same year established the basis for rural local government (20 years later it morphed into the governance of small cities). A general municipal franchise ordinance was promulgated by the Mandatory rule in 1926. It extended municipal voting rights to resident male tenants even if they held no property provided that they paid at least one Palestinian pound in municipal rates annually (Elazar and Kalchheim 1988).

It was not until 1934 that a comprehensive Municipal Corporations Ordinance was passed. It empowered the high commissioner to establish new municipalities or change the boundaries of existing ones on the recommendation of a public committee of inquiry, a system which continued after the establishment of the State of Israel, with the Minister of Interior inheriting the powers of the high commissioner. The Ordinance detailed the method of elections, duties and powers of council members and the municipality, revenue sources, procedures for approving the municipal budget, methods of financial control, and regulations for filling the major administrative positions of town clerk, treasurer, and medical officer. The law also established procedures for council and committee meetings and rules for establishing committees.

Local councils were empowered to enact bylaws but only on those subjects listed in the ordinance, and subject to the high commissioner's approval. The district commissioners were given oversight and budgetary approval powers, while the high commissioner retained the right established during the Ottoman period to nominate the mayors and deputy mayors. This Ordinance was carried over after 1948 and many of its provisions still apply although they continue to be modified.

In 1941 a new ordinance came into effect, providing for the establishment of regional councils based on patterns of governance developed for existing federations of Jewish settlements, and formalizing the arrangements which had already been developed in the Jewish sector.

In 1945 the Local Authorities (Business Tax) Ordinance was promulgated which allowed local authorities to enact bylaws to tax businesses operating within their boundaries, subject to approval by the high commissioner. This completed the bundle of local legislation under the British Mandate. The Arab small towns and village residents regarded this ordinance as interference with their traditional way of life and opposed its implementation. Instead, Arabs secured a Village Administration Ordinance enacted in 1944 which provided for a more modest change in the traditional government of the villages. By the end of the Mandate in 1948 there were only 11 Arab local councils, while the number of Jewish local councils had increased to 26, in addition to four regional councils. From the beginning of the 1930s until the establishment of Israel, the local government was set up and operated under foreign central government and within an ethno-national and geopolitical conflict which had direct impact on the local authorities' work. The conflict continued to affect the Arab local authorities after the Israeli state was established.

With the establishment of the State of Israel in 1948, local government left the center of the political stage. The new state began to assume responsibility for many public functions which had rested in local government's hands for lack of central institutions. At the same time, the mass immigration to Israel in the years between 1948 and 1953 shifted the patterns of settlement in the country in such a way that the rural Jewish localities which possessed the best access to the state and the most power to maintain their local autonomy declined in importance relative to other local communities. On the other hand, the development of Arab towns whose populations were expelled or fled in 1948 such as Tiberias, Lod, and Ramla, and the immigrant settlements, potentially the least powerful local communities, became significant elements in the constellation of local governments. In this situation these now predominantly Jewish localities began to absorb new immigrants in what were called development towns. These towns are located in peripheral areas and follow a policy of population dispersal. They still suffer today from socioeconomic hardships.

The new State of Israel took upon itself to foster local authorities' institutions. Reversing the pattern established in Mandatory days, the state authorities themselves moved to establish new local authorities. The number of Jewish settlements enjoying municipal status rose from 36 in 1948 to 107 by 1968 and to 148 by 2010. The number of regional councils rose from 4 in 1948 to 55 in 2013 (CBS 2014; 154–155). Moreover, new rural settlements were encouraged to develop local committees of their own for internal self-government. Arab villages were also encouraged to establish modern municipal governments and did so in substantial numbers.

In the late 1950s, the tide began to turn as local governments began to find their place in the framework of a state in which power was divided on other than territorial bases. The process of adjustment which began at that time is not yet complete. In the case of government services, after the period of mass transfer of functions to the state, the country entered a period in which shared or cooperative activity began to be emphasized. With regard to functions defined as state services, the state took primary responsibility for program initiation, policy making, and finance, while program administration—the actual delivery of services—was increasingly transferred to local government. In cases where the division was not so clear-cut, responsibility for the delivery of services was somehow divided between the state and the localities. This became true over a wide range of functions, from welfare to education, civil defense, and sewage disposal.

Moreover, local governments in Israel undertake a range of social and cultural functions which extend beyond the ordinary police functions of local government. These range from the provision of religious services to the management of orchestras and drama groups, from the maintenance of day-care centers to the awarding of literary prizes.

Role and Duties of Local Authorities

Local authorities played and continue to play an important role in the diversity of Israeli society, where most of Jewish society were immigrants (Nachmias 2006). To promote new state building, the local authority played a significant role in societal integration and population dispersal (Elazar and Kalchheim 1988). The tasks of the local government still encompass the following four points in order to implement state goals and meet societal needs:

* The provision and administration of governmental services
* The recruitment and advancement of political leadership
* The fostering of channels of political communication between the governors and the governed
* Maintaining a necessary or desired diversity within a small country where there are heavy pressures toward homogeneity

All four of these tasks are of significant importance in the integration of what is still a very new society of immigrants or the children of immigrants. The role played by local government in meeting these challenges makes it a far more vital factor in the Israeli context than is often credited.

The local authority provides its residents, commercial firms, and other institutions within its jurisdiction with a wide range of services. It develops its physical infrastructure, road system, water supply, refuse collection and disposal systems, sewage systems, and parks. It is responsible for environmental protection (public health, nuisances, cleanliness, etc.) and, along with the Ministry of Education, culture and recreation and the education system. The local authority builds schools and is responsible for providing equipment and their maintenance. Prekindergartens and secondary education institutions are established and administered by the local authority, but some of these facilities may be owned by nonprofit organizations with aid given by the local authority. Local authorities also promote and financially assist cultural and sports activities (libraries, museums, youth clubs, etc.) and some maintain orchestras, choirs, theaters, and similar enterprises. The local authority provides social welfare services, with its social workers taking care of families in need, as well as distinctive groups such as the elderly, and people with physical and cognitive disabilities, drug addictions, and other special needs.

The local council has an important role in town planning. The Planning and Building Law, 5725–1965, sets forth the principles according to which town planning is undertaken, as well as planning institutions that must act at the local, district, and countrywide geographic and administrative levels. The law grants the local planning commission considerable independence while also expanding the regional and countrywide dimensions of planning. The local planning commission draws from members of the local council. The law gives the local planning commission responsibility for day-to-day management and on-site compliance with regulations.

The district planning commission is a joint state-local authority comprising representatives of those government ministries whose field of action is relevant to the issues of planning, and representatives of the local authorities in the region. The district commission approves detailed local plans and also acts as a forum for handling appeals of decisions made by local commissions.

The local authorities are empowered to levy local taxes and impose and collect various payments for services and concessions (Khamaisi 2002). In general, they have the powers and means to manage finances. They prepare their own budgets, which then must be approved by the Ministry of Interior. Each municipality must employ an auditor to check its activities.

Ben-Elia (2004) summarizes the development of the local government in Israel by dividing it into three stages and makes suggestions for preparing local government for stage four. The first stage was to craft a new public agenda that shifted the traditional dominance in Israel of security-related issues toward social and civic concerns. The second stage exhausted patterns of public governance and defined the need for institutional change. In the last two decades, local government played a key societal role. The third stage witnessed a faltering, ineffective central government and increasingly pressing local problems which encouraged the emergence of a more assertive, entrepreneurial, and effective local governance. A de facto institutional decentralization expanded the role of local government, transforming it into a multipurpose entity handling a widespread range of critical services and activities.

After 20 years, local government has reached a critical impasse. Contradictions in ongoing policies and ideological changes have exposed the limitations of this present system of local government. The local, national, and international economic and social changes require that local government prepare itself for the challenges facing new generations. Ben-Elia articulates his view of what the new local government in Israel should look like as going beyond the immediate problem of finding technical solutions to a system on the verge of collapse. A central argument of Ben-Elia is the emergence of a new social and civic agenda that demands improved governing capabilities and new guiding values.

Local government in Israel has transitioned from its role as the representative of the local community to a provider of services which initiates development and functions as a 'business' in accordance with economic efficacy to maximize its benefits. While local government in Israel has experienced several reforms, these were mostly small and incremental. Razin (2003a, b) analyzes the barriers that have so far hampered attempts to reform Israel's local government. These barriers include ethno-religious heterogeneity, sociopolitical fragmentation, an increasing role of courts as the arenas for societal conflicts, and the significant role of national parties in local politics through multiparty coalition government structures. Razin contends that partial and flexible implementation of control and monitoring mechanisms entails problems of fairness and proper administration but might also help to prevent collapse of the system. This crisis of local government could open a window of opportunity for change. Several stakeholders call for wide-scale reform of the local authorities in Israel, including ratifying a new local government law.

Organizing Local Authorities into a Union

Local government is organized into a Union of Local Authorities in Israel (ULAI), which is the umbrella organization of 265 Israeli local authorities that represents them before the national government in dealing with their daily issues and problems. ULAI was established during the British Mandate in 1938 as the "The League of Local Councils." The main role of ULAI is to oversee the shared interests of all local authorities and represent them in their affairs with the government and its ministries, the Knesset (Parliament), the Jewish Agency, Histadrut (General Federation of Trade Unions), and other institutions and municipal public organizations. Furthermore, the ULAI advises local authorities regarding a range of municipal matters including education, welfare, economy, water, security, status of women, labor relations, law, and the courts (see http://www.masham.org.il/English/Pages/AboutUS.aspx).

In recent years the ULAI has become a leading force within the State of Israel through involvement in areas of civil life, helping local authorities to provide the necessary level of municipal services to their residents especially in the fields of education and welfare.

A Crisis Situation

Local authorities in Israel, especially the poor and marginalized ones such as the Arab local authorities, are at an impasse. This crisis has become a permanent problem with which local authorities are unable to deal on their own, without structural change and an increased injection of external resources as argued by Ben-Elia (2006). For the purposes of our discussion, the crisis situation that led to this has the following characteristics:

- Little or no ability by a local authority, as a level of government, to govern, make decisions, and implement its decisions as a representative governmental body. This increases the "democratic deficit" (Moravcsik 2004), in a process that develops in tandem with the emergence and growth of the fiscal deficit. The democratic deficit, caused by this crisis of governance, is characterized by dysfunctional local authorities, a problematic decision-making process, and a low level of services provided to citizens. Functional, managerial, and behavioral characteristics of the local authority impede relationships with central government and increase this deficit (Hasson 2006).

- A local authority's inability to fulfill its statutory obligations and to exercise the powers delegated to it by the central government as part of the decentralization policy. This policy is not well defined and does not clearly stipulate local powers, failing to clarify the division of responsibilities between central government and local government within the latter's jurisdiction, types of activities that the local authority is charged to plan and implement, and services it is supposed to provide. Local authorities have obligations to their citizens and those working in their jurisdictions; in many cases, however, they do not meet these obligations and find themselves in violation of the law. Consequently these authorities are exposed to fierce criticism by the courts, the central government, the state comptroller, and public opinion. This crisis situation harms both citizens and vendors who provide services to local authorities. When the Union of Local Authorities comes to the defense of some local authorities, attempting to protect them against creditors or lobbying for laws that would grant them immunity from lawsuits and relieve elected and appointed officials of responsibility, it undermines public confidence in the local authorities and amplifies the problems.
- The failure to provide basic services (education, culture, and community services) that meet residents' needs and expectations, or their irregular or insufficient availability or poor quality, and a shortage or faulty maintenance of infrastructure (water, sewers, drainage, roads, electricity) are other signs of crisis.
- Another indicator is the lack of an economic capacity to fund the regular provision of basic services according to a fixed, transparent, and clear schedule, and the inability to meet financial obligations (workers' salaries, debt repayment, etc.). A crisis, by this definition, is a situation in which the local authority has insufficient fiscal resources to function and fund its activities. There are several causes for this, including the cutback in regular transfers from central to local government, reduction in the budgets transferred as balancing grants, and reduced (or nonexistent) locally generated revenues.

The crisis of local government in Israel is not unique to the Arab local authorities, nor is it unique to Israel. It is a widespread international problem for which there exists accumulated experience for solutions by means of territorial, functional, and administrative reforms, or by the revamping of the system, including redefinition of the relationship between central and local governments and between government and citizens.

The planning process in Israel is predominantly rational, incremental, and top-down. Although the Planning and Building Law of 1965 which was based on the British Mandate Town Planning Act of 1936 has since been modified, it continues to resemble closely the 1936 Mandate Act whose overriding principles are centralization and hierarchy. When the 1965 Law was amended to allow more local power, as the foregoing synopsis described, many localities proved ill equipped to take advantage of their additional powers, both because their independent financial bases suffer from limited taxing powers and lack of adequate professional staff to exercise them (Ben-Elia 2004; Shmueli 2005). All building and development requires licensing by the Local Planning and Building Commission. Permits are issued if they conform to approved local Outline or Detailed Schemes (akin to statutory master and site plans) which must be congruent with District and National Outline Authorized Schemes according to the Israeli Planning and Building Law of 1965 and its amendments.

Procuring building permits is one of the major lacunae for the Bedouin in the Negev. There are three dimensions to issuing permits. The first deals with land zoning and ownership with the precondition of having an approved outline plan which regulates plot parcellation. Then documentation as to ownership or leasing is required. Thus, even after a village receives formal, government recognition and an approved master plan, if there is a legal dispute (a land claim) over land ownership, the government, represented by the Israel Land Administration, will not approve the issuance of building permits. The ILA will not allow Bedouin with land claims to sign the application as land owners, and Bedouin are not willing to sign as leasers.

The second dimension is financial. Building permits are accompanied by high costs of land parcellation in accordance with the official plan, costs of leasing or buying land (and accompanying taxes), and paying fees for land betterment (infrastructure, etc.). Without government aid, the Bedouin do not have the funds necessary for projects of this magnitude. And the last dimension is related to sociocultural customs. The speed required by the government to build permanent dwellings is quite unlike the slow building of a home according to financial ability, which expands over generations to accommodate growing families. So whereas the need for recognized permanent housing and infrastructure will improve the very basic rights which are missing—it will also create a financial burden for economically weak Bedouin families. These are some of the problems and incongruities which result in lack of development in most recognized villages, making them indiscernible from the unrecognized.

Planning takes place within the system of municipal organizations which serve as managerial frameworks for the affairs of the population residing within a given community (Razin 2003a, b; Barlow and Wastl-Walter 2004; Dollery et al. 2005; Lazin et al. 2007). In Israel, these frameworks are organized within a hierarchy of settlement types based on population size: cities are urban forms of dense settlement; local councils govern towns and some larger communities (villages); and regional councils serve a number of small- to mid-sized villages located in suburban, exurban, and peripheral settings (Appelbaum and Hazan 2005; Ben-Elia 2004; Razin 2003a, b; Razin and Hasson 1992; Razin et al. 1994). Local government officials are initially appointed by central government, and subsequently elected by local residents. The process of municipalization in Israel enjoys the support of central government, which over the past two decades has encouraged decentralization of governance (Nachmias 2006; Razin 2003a, b; Ben-Elia 2004; Efrati et al. 2004). This has not been fully applied, however, to the Negev Bedouin communities.

Central government retains a number of mechanisms enabling it to perpetuate its differentiated control over local government, thereby maintaining the decision-making power over the geographical jurisdictions allocated to each authority (Meir 1999). Thus, the government may establish a regional council whose constituent communities lack territorial contiguity or which cannot expand its jurisdictional boundaries, thereby creating obstacles to efficient economic and social development.

The Bedouin Abu Basma Regional Council and its new offshoots—Neve Midbar and Al Kasum—are such examples.

Unrecognized villages have no official outline plans or building permits and lack basic infrastructure. Some of them will probably be recognized as part of a planning process that will include them in a National Outline Plan (35) and in a partial District Outline Plan for metropolitan Beer Sheva (23/14/4). For communities excluded from this framework, there is particular need for a complementary planning process leading to their recognition in keeping with the recommendations of the Goldberg Committee, to be expanded upon in Chap. 7.

The community structure of any given populace, including its stratification in social affiliation, class, and ethnicity, has a direct impact upon municipal organization. This is especially pronounced in the case of a traditional, patriarchal society such as the Bedouin that continues to cling to its social frameworks, separating it from other segments of society. Lack of social mobility reinforces territorial enclosure and impedes cooperation and interaction with external groups. Although Bedouin society is no longer nomadic, those living in villages remain strongly based upon farming and herding, and therefore need access to land.

Social stratification–tribal tradition, social class, land ownership, internal evictees-refugees, and subordination of women (Marx 1980; Falah 1989; Meir 1997; Porat 2009) reflects the complexity of achieving consensus within the Bedouin community and the heightened urgency of exploring a flexible set of planning options for creating new or revised municipal structures.

Chapter 7
Resettlement Planning 1948–Present

Background to the Planning Process for Bedouin Towns and Villages

Modern municipal management structures among the Arab population in general, and the Bedouin population of the Negev in particular, are recent. Beer Sheva, founded in 1906, was only established as a local authority in 1926 (later becoming the regional city for the South) and was the first municipal framework in the Negev.

The Bedouin community's tradition-based legal framework includes mechanisms for managing society in accordance with rules and customs that have emerged from Islamic religious law, shaping the culture and governing the relations between individuals and groups. In addition, each community maintains arrangements based on its own customs and traditions (Al-Haj and Rosenfeld 1990). The settlements are managed by the tribal leaders—mukhtars and sheikhs.

The Israeli planning paradigm for the Bedouin has historically been based on territorial considerations. Most government planning policies have aimed at concentrating the Bedouin in high-density towns with defined and limited jurisdictions, reserving inadequate land for public use and economic development (Meir and Marx 2005). These policies can be divided into three main periods. The first, 'regional concentration,' occurred in 1949, immediately after the establishment of the State of Israel. For geopolitical military considerations, the Israeli Government relocated the remaining Negev Bedouin to the Syag region, where most remain, although today some of the towns and recognized villages extend beyond this boundary (see Figs. 7.1 and 7.2 later in this chapter). They received neither housing nor compensation. A military governor was appointed to control the Bedouin and they were not allowed to move outside the Syag region without a permit. The relocated tribes often settled close to tribes already living in the area with whom they had some social/kinship relationship. The Syag thus became home to two groups of Bedouin, those who lived there prior to the 1949 and continued to live on their lands

© Springer International Publishing Switzerland 2015
D.F. Shmueli, R. Khamaisi, *Israel's Invisible Negev Bedouin*,
SpringerBriefs in Geography, DOI 10.1007/978-3-319-16820-3_7

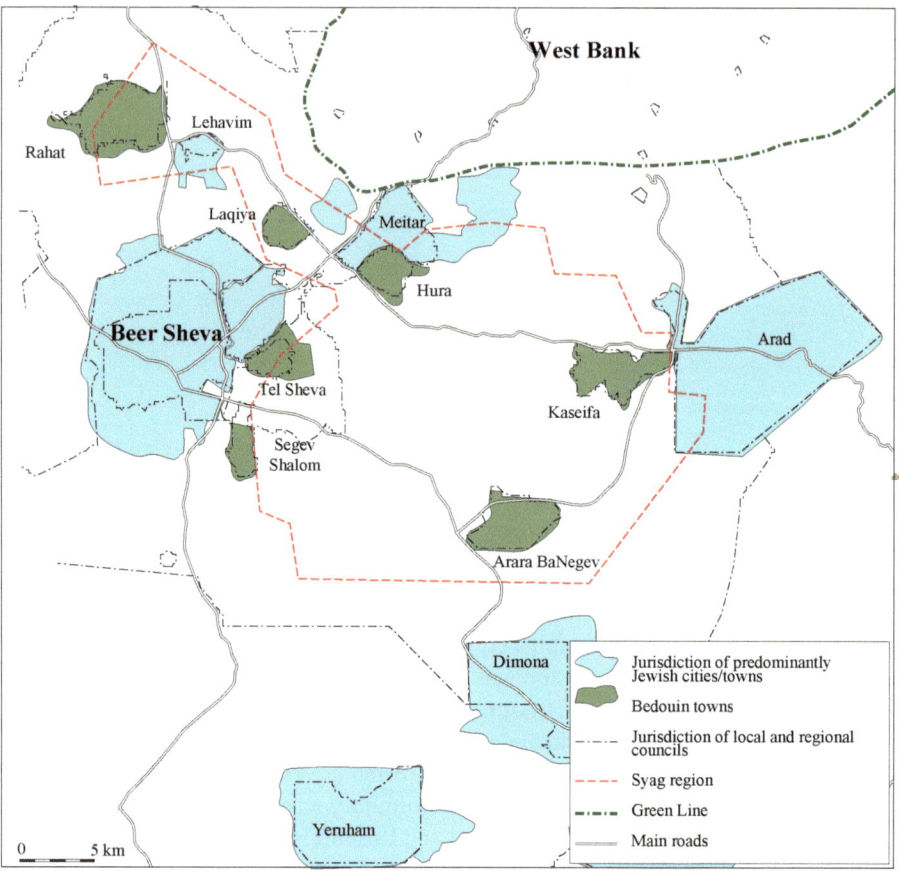

Fig. 7.1 Geographical distribution of the seven Negev Bedouin towns

and territory, and internal refugees who were removed from their organic territory and relocated (Shmueli 1980).

The Bedouin population grew rapidly (3.7 % per annum, CBS 2014) as a result of high birthrates, reinforced by polygamous practices, spreading itself throughout the Syag. The government's next concern was how to limit the Bedouin territory within the Syag. This second planning period can be referred to as 'local-urban concentration'. It began in 1964, when the government decided to build Tel Sheva, the first Bedouin new town. The planning model was fashioned on that for new Jewish neighborhoods—small plots (less than 500 square meters) on which houses of 50 square meters were constructed. This model ignored the sociocultural and land needs of the Bedouin, whose families were much larger than those of Jewish immigrants and who required expanded plots for maintaining even a handful of animals. It was a failure.

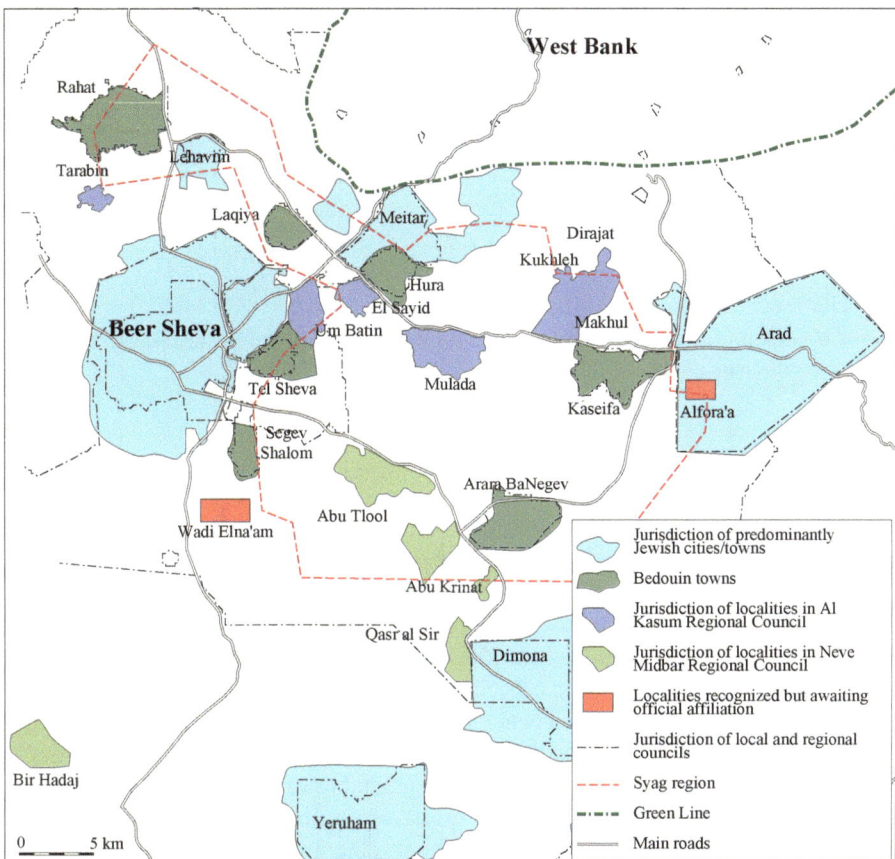

Fig. 7.2 Regional councils

The government tried to improve this model and in 1970 built the second Bedouin town of Rahat. Here nuclear family houses were placed on larger plots of 1–2 dunam and organized spatially in neighborhoods according to clan and tribal affiliations. Rahat was divided into 36 clan neighborhoods. The government began to push and promote the internal refugees and Hummran peasants in the Syag to settle and reside in the first neighborhoods. This experience was generally viewed by both the government and the Bedouin as being somewhat more successful than Tel Sheva, and this improved 'local-urban concentration' model was followed in the remaining five Bedouin new towns (Segev Shalom, Arara BaNegev, Kaseifa, Hura, and Laqiya) with the ultimate aim of concentrating the entire Bedouin population of the Syag among the seven planned new towns (Fig. 7.1; Table 7.1). While at the time only one planning model was developed for the Bedouin, diverse options existed for other segments of Israel's populace from development towns to small rural villages.

Table 7.1 Population growth of the planned Bedouin towns, according to Formal Register Data among the Bedouin in Negev between 1948 (1983 establishment of first towns) and 2013

Year/name of town	1983	1995	2008	2012	2013
Rahat	9.2	23.2	50.0	56.9	58.7
Laqiya			9.2	10.2	10.8
Hura		2.1	16.0	17.5	18.3
Segev Shalom		2.4	7.0	8.2	8.5
Tel Sheva	2.5	6.9	14.6	17.0	17.5
Kaseifa		4.3	16.2	17.5	18.1
Arara BaNegev	1.3	7.4	11.5	14.2	14.8

Source: Central Bureau of Statistics, 2014, Statistical Abstract of Israel No. 65, Central Bureau of Statistics, Jerusalem

This paradigm took into account only one aspect of Bedouin life—tribal affiliation. The planning policy was to maintain the hierarchical social stratification reinforced by custom and Bedouin law which motivate the sons of the same clan/kinship to live in the same territory and the same neighborhood. However the government and planners ignored the question of land rights of the more sedentary 'original' Bedouin, and did not sufficiently take into account the impact of their sedentarization which gave greater scope to individual behavior. Both sets of Bedouin who took up residence in the seven towns, especially the older generation, underwent a rapid urbanization process with limited benefits of urbanism and began to feel under siege.

These towns were planned as 'bedroom' or 'satellite' localities to Israel's southern city of Beer Sheva. However, they are unsuitable to the rural character of the population, lack an economic base, and offer very limited employment opportunities, which, alongside the rapid natural demographic growth and rigid social structure, create huge needs which are not met by municipal and governmental institutions. Discriminatory employment practices and peripheral location contributed to their deterioration into Israel's most impoverished towns, wracked by unemployment, youth crime, theft, property damage, prostitution, etc. (Lithwick 2002). Large numbers of the Bedouin refuse to move to the towns, particularly clans and families who have land claims within the Syag. These clans and families continue to grow and expand in their organic locations, and the government and the planning and municipal systems continue not to recognize them. By 1995, the numbers of unrecognized villages in the Negev had reached 45, with varying population sizes and spatial distributions. During this period, the sedentarization of the Bedouin in officially recognized municipalities, as well as in illegal settlements, was monitored by a paramilitary 'green' patrol established by Ariel Sharon in 1976. The patrol had the mandate to pull down illegal Bedouin tents, control herd sizes and grazing areas, destroy crops which lacked governmental permission, impose fines, and evict inhabitants (Koeller 2006).

In the struggle by Negev Bedouin to secure recognition for their unauthorized settlements and acquire basic services then denied, the Regional Council for the Unrecognized Villages of the Negev (RCUV) was established in 1996 as an

Association to represent local committees in their campaign for recognition. Prior to the establishment of the RCUV, two NGOs—the Association for Support and Defense of Bedouin Rights in Israel (founded and headed by Nuri Ul-Oqbe in the 1970s) and the Forty Association (established in 1988 to advocate recognition of Arab villages in the Galilee and Negev)—fought for recognition of Bedouin villages, to stop urban resettlement, housing demolitions, and the government confiscation of private lands. Both associations prepared alternative and plans for recognizing new Bedouin villages in the Galilee (Sternberg Ben-Baruch 2013), and in the Negev (Khamaisi 1990). Most of the villages in the North which were outlined in these early alternative plans were recognized by the government in the late 1980s and 1990s, and some of those in the Negev have received recognition since 2003. The RCUV, although not recognized by state agencies, initiated preparation of a master plan for all unauthorized settlements, working with other bodies through litigation and contacts with the planning authorities to achieve its goals (Yahodkin 2004).

The third planning period began in the late 1990s and includes two models. The first, a continuation of stage two, consisted of re-planning within the seven Bedouin towns. The second model focused on the planning of localities/villages recognized at the beginning of 2000. These villages were organized municipally into one regional council—Abu Basma which was founded at the end of 2003 and functioned until 2012 when it was split into two regional councils, Neve Midbar and Al Kasum. Abu Basma included new Bedouin communities as they were recognized and by 2012 comprised 11 communities.

Although 30,300 people made up the jurisdiction of Abu Basma, in 2012 only a small number were registered with the Ministry of Interior. The numbers differ: as of May 1, 2012, only 6,539 residents were registered in the villages (see Table 7.2) according to the Population Administration of the Ministry of Interior, while the Central Bureau of Statistics had 13,300 registered in 2012 (Table 7.3). Whatever the

Table 7.2 Abu Basma communities at the beginning of 2012

Name of community	Area in Dunam	Estimation population living in the area and get some services	Population registered with the Ministry of Interior
Um Batin	6,600	2,900	2,134
El Sayed	3,100	3,000	603
Dirajat	1,000	850	727
Kukhleh	300	250	70
Mulada	11,100	3,500	213
Makhul	15,000	3,000	231
Tarabin	1,100	1,000	383
Abu Krinat	7,300	3,600	292
Abu Tlool	11,500	5,000	14
Bir Hadaj	6,500	4,200	734
Qasr al Sir	4,800	3,000	1,149

Source: Area calculated from govmap-Center of Israeli Mapping, http://www.govmap.gov.il, registered population from the Population Administration, Ministry of Interior http://www.cbs.gov.il/ishuvim/ishuvim_main.htm

Table 7.3 Registered population of recognized villages within regional councils (Abu Basma was split into Al Kasum and Neve Midbar Regional Councils in 2012) according to Formal Register Data growth among the Bedouin in Negev between 2008 and 2013

Year/name of regional council	2008	2012	2013
Abu Basma	11.2	13.3	
Al Kasum		7.6	8.2
Neve Midbar		5.5	6.9

Source: Central Bureau of Statistics, 2014, Statistical Abstract of Israel No. 65, Central Bureau of Statistics, Jerusalem, Table 2.22, page 155

correct number is, only a small proportion of people living in the recognized villages have registered as such, primarily due to the Bedouin fear of negating land claims by registration, need to pay taxes which they cannot afford, and lack of trust in national government. The Council included the growing number of recognized communities within its jurisdiction, serving as a platform for incorporation of villages as they received governmental recognition and was responsible for education and welfare of the entire Bedouin diaspora, including those residing in unrecognized communities. As can be seen from Fig. 7.2, there is no territorial continuity between the communities of the regional council. The Council is economically weak without the ability to provide funding for the post-planning development of the communities, and the problem of land claims prevents the progression from approved plans to implementation in lieu of building permits. This contributes to a deepening of the dispute between the Bedouin population and the state, and further impairs the effectiveness of this municipal structure.

In 2012 Abu Basma was split into two regional councils—Neve Midbar, incorporating the Southern communities of Abu Krinat, Abu Tlool, Bir Hadaj, and Qasr al Sir with 15,800 residents and with responsibility for the education and welfare of the unrecognized villages; and Al Kasum including the northern communities of Um Batin, El Sayed, Dirajat, Kukhleh, Mulada, Makhul, and Tarabein with 14,500 residents. The reasons for the split were twofold: Abu Basma was already among the largest of Israel's regional councils and the number of residents was expected to double by 2030. By law new regional councils have appointed heads and must hold democratic elections after 4 years. In Abu Basma the government had extended this period to 9 years but elections were slated to be held at the end of January 2012. Among the recognized villages, almost all petitioned against elections on tribal clan grounds—rejecting the idea of a head being elected from a larger tribe (or one with more registered voters). Splitting the Abu Basma Regional Council would automatically mean another four years of appointed heads for the two new councils.

The two new councils still face hurdles:

- Lack of territorial continuity.
- The geographical locations of unrecognized villages which are viable for recognition and their social structure would still render the two councils unwieldy.
- Spatial planning.

- Social-cultural structure—clans and community.
- Economic development.
- Adequate service packages for residents.
- Should these councils grow without splitting again there is a potential of friction and conflict among the communities themselves.
- Weak local councils and lack of participatory governance.
- Lack of municipal facilities.
- Lack of economies of scale for shared infrastructure and services.

This model gives greater consideration to the sociocultural imperatives of tribal separatism, land ownership, and existing building structures. However it is still guided by territorial planning objectives whose aim is rapid urbanization and modernization among the Bedouin, ignoring the land needs of rural development which the Bedouin continue to request. Many of the plans for the individual villages still lack a sufficient economic base for development. Some plans do include agricultural development such as in Tarabin, and an industrial zone such as in Um Batin. A conflict assessment published at the end of 2006 (a year and a half after recognition) reflects the frustration of Um Batin residents over the slow pace of implementation (Consensus Building Institute 2006). A master plan for Um Batin has since (May 15, 2011) been approved but no building permits have been issued to date (2014).

The planning principles incorporated into the village of Abu Tlool recognize the existing organic buildings and the sociocultural constraints within the village, as well as encourage public participation, making provision for managing the conflict between the Bedouin plans and the national and regional plans which limit local development. Although passed in 2014, here too, no implementation has occurred.

The changes in the planning approaches to Bedouin settlement have been excruciatingly slow—almost too late. They are a response to Bedouin pressure for involvement in civic society and a strengthened awareness of their own needs and for more democratization of the planning system. This recognizes the failure of the planning and development of the Bedouin towns, and the need for planners to respond and be more sensitive to cultural and social needs and the agricultural orientation of some of the villages. The territorial considerations and unaddressed land disputes continue to shadow Bedouin rights and planning resolutions. As a result of this long history of government discrimination and ignoring Bedouin land rights and community needs, the Bedouin continue to respond to planning with mistrust and as a government tool to limit their development.

Government Policies and Responses

Since Israel's statehood in 1948, the question of Negev Bedouin resettlement has been on the government and public agendas. The contradictory interests and territorial needs of the Israeli Government (as manifested in policies and actions) and those of the Bedouin community have evolved. The transformations over the last six decades have changed from the government treating the Bedouin as invisible or temporary citizens to visible citizens struggling for equality and equity, and the Bedouin transformation from weak reactive responses to active responses, initiatives, and demands reflecting their needs and interests. The relations between governmental agencies (national and district) on one side and the Bedouin communities and NGOs on the other can be described as a tug of war. This game began in the early 1950s and continues today, using a strategy of checks and balances, despite asymmetrical power and roles. This period from 1948 to the present is one of planning and development of the Negev area in general including establishing and developing Jewish settlements, national infrastructure, and military bases, and dealing with issues of the Bedouin in particular—and the barriers they present to reaching Israel's regional planning goals. It is a mistake to relate to the Bedouin resettlement and spatial reorganization policies as simply an ethno-national issue; rather, it has to be examined in the wider context of the Negev region's role in national geopolitics.

The two sides in this tug of war have both contradictory and common interests. The government's guiding principles are both territorial and functional, resulting in its goals of decreasing the area in which the Bedouin live or herd, transferring the land to state land, strengthening the Jewish regional presence, modernizing and urbanizing the Bedouin communities, transferring national infrastructures to the Negev—the least populated region in the country—and saving natural reserves elsewhere. The goals of the Bedouin include preserving their land and rural existence; safeguarding their traditions, culture, and social structures; gaining services and living conditions equal to those of other Israelis; and using their demands as leverage for future Negev development. Although both share development as a common interest, development has different meanings to each side. The government, in seeking to provide services, sees development as Bedouin resettlement to concentrated urban modern localities and a shift in the their economic base with increased opportunities which would follow greater concentration. The Bedouin see development as a way to gain municipal services and economic opportunities in dispersed, more rural settings and secure their land claims ensuring their future in the region. These opposing interests enveloping the government-Bedouin dispute exist against the background of the national and regional Arab Israeli conflict—a geopolitical and ethno-cultural power struggle—and is affected by it both directly and indirectly. Misunderstanding and mistrust are abundant on both sides.

The dispute has undergone numerous transformations: from small and local to big and regional; from regional and large territorial expanse to local concentration in the Syag; from military to civil governance; from a Bedouin discourse focused on survival to demands for equal quality of life; from a traditional economic existence to a more diverse modern economic structure; from separation—"far from the eye far from the heart"—to societal integration and an understanding of and participation in the rules of the game, as set by government procedures; from uniformity of problems facing the Bedouin community to complexity and a diversity of needs; and from political marginalization of Bedouin issues to these issues becoming central and national in scope engaging most government agencies on the national, regional, and local levels, the judicial system, and NGOs.

These transformations were both internal and external to the Bedouin community. The internal transformations can be summarized thus:

- Demography: Rapid population growth, from approximately 13,000 inhabitants in 1951 to 224,000 citizens in 2013.
- Settlements and municipal structure: Establishment of planned urban localities, the first, Tel Sheva was established in 1964 and populated in 1968. Today there are 18 planned and recognized localities, managed by 9 different local authorities (7 townships, and 2 regional councils together managing 11 recognized villages). About 75 % of the Negev Bedouin reside in these communities.
- Social structure: Mainly organized around tribe, clan, and kinship in addition to locality/village affiliations. Today social structure is also developing around needs, interests, and issues. The social structure remains strongly tribally affiliated; however individualism and nuclear family are gaining in importance.
- Democratization: Participation in municipal and national elections. There is still a deficit in democratic behavior with tribal affiliations foremost. However there is an understanding of democratic structure and Bedouin do participate in national political parties.
- Territory and land attachment: Some Bedouin do not have land claims; others live on land they claim their own designated as Mawat by the state. Others suffer from land confiscation. Some have recognized Ottoman or British land deeds. Among the Bedouin groups, one, for instance, wants to expand its locality's jurisdiction, a second claims land/territorial ownership, a third demands recognition of its village in situ or allocation of other state-owned land for development, and yet a fourth asks the government to return to its land which was confiscated. Thus, among the Negev Bedouin territorial considerations hold a myriad of meanings and implications with regard to land claims, fragmentation of land among various inheritors, need for jurisdictional expansion, or reorganization of jurisdictional oversight—in addition to government recognition, planning, and perhaps aggregation of new, currently unrecognized villages.
- Awareness: The youth are educated, aware of both their own needs and those of their community, and their rights. They aspire to participate in the production of their space and environment, and have introduced a new discourse and actions. This has led to the establishment of NGOs dedicated to Negev Bedouin issues, cooperation with national civic society, use of the legal system to gain rights and services, and the political system to leverage stakeholders.
- Emergence of a middle class: Changes in economic structure, benefits from economic opportunity and prosperity in Israel as a whole including the Beer Sheva area, and an increasingly educated generation have led to the emergence of a middle class among the Bedouin. This class is aware of its role as leaders of change.

External transformations include the following:

- The national plan for the Negev: Israel sees the Negev as a potential reservoir for national infrastructure development and Jewish settlement. The concentration of Bedouin in a smaller territory and transferring land ownership to the state serve the development of such projects as military camps, polluting industries which are difficult to locate in more densely populated areas, airports, and preservation of desert natural reserves. Expansion of Jewish settlement in the region is an integral part of the policy.
- Fear of territorial continuity of Arab Palestinian settlements: Bedouin territorial dispersal in areas close to the West Bank and the Gaza Strip creates the threat of Arab Palestinian continuity between the Gaza Strip and the West Bank through the area currently occupied by dispersed Bedouin villages.
- Democratization and prosperity: Despite the civic discrimination and territorial, planning, and development restrictions from which the Bedouin suffer, they benefit from the democratization process in the country, from the different views, interests, and attitudes of the various government agencies and district bodies. Bureaucratic complexity and changing detailed policy and actions toward the Bedouin have contributed to creating more varied solutions regarding the planning and recognition of Bedouin villages. The Bedouin also benefit from Israel's economic growth and prosperity.
- Over-legalization: The judicial system has been used by the government agencies to impose decisions on the Bedouin—Mawat, transfer to state land, enforcing building laws, and demolishing homes. At the same time the Bedouin use this legal system to protect themselves and to obtain services such as education and water, as well as some of their advocacy objectives.
- International climate and support: Israel wants to be part of the modern liberal and democratic international community. The international community's awareness and concern for the rights of indigenous people are strong, particularly in the world of open media. The Bedouin are leveraging this support and the Israeli Government is influenced by international opinion.
- Political power: The Israeli political system is strongly influenced by stakeholder demands and lobbies. Bedouin involvement in political parties and the establishment of dedicated NGOs have impacted both policies and actions among and toward the Bedouin.

These internal and external changes affect the territorial and functional governmental policy for and among the Bedouin. Since the 1950s numerous public, governmental, and professional committees have been established to deal with the issue of resettling the Negev Arab Bedouin and suggesting solutions for the land dispute between them and the state. These committees submitted their recommendations and suggestions to government; reports analyzing the committee's findings have been published (for instance, Dor-On 2009; Negev Coexistence Forum for Civil Equality 2009) but to date no agreement has been reached. Table 7.4 which follows summarizes the main government policy benchmarks and Bedouin reactions since 1948.

More detail as to some of the major points summarized in the table follows:

Table 7.4 Government resettlement policies and Bedouin response 1948–2014

Year	Governmental action	Bedouin reaction
1948	Israeli army occupies the Naqab/Negev desert, including areas designated by the United Nation's Partition Plan, resolution 181 for the Arab State	Fight against the occupation but lose the war. Majority of the Bedouin go into exile (approximately 85,000 leave and 13,000 remain)
Period 1: Regional concentration		
1949	A military governor is appointed to oversee the remaining Bedouin. He moves them into the concentrate Syag region, most of which had been designated a part of the UN-proposed Arab State (resolution 181)	Some of the tribes/clans oppose the transfer to the Syag; however they do move, are settled on state (Israeli) land, some are told that this is a temporary condition which will be reversed. This becomes a permanent solution, creating a new internal refugee situation for the Bedouin
1954	Israeli Government grants the Bedouin identity cards and citizenship (according to the Citizenship Law 1952); registers each individual's address according to tribal affiliation, not recognizing their geographic location as permanent villages	Bedouin accept citizenship, and many volunteer to serve in the Israeli Defense Forces. They request recognition of their villages. A few request permission to move back to their original land and territory and some requests are approved by the military governor. A few, mostly those without land attachment, move to Israel's northern or central districts
1963	The Ministry of Agriculture, headed by Moshe Dayan, develops a policy which would concentrate the Bedouin in few villages in the Negev and supports transfer or out migration of about two-thirds of the Bedouin to the central district, in towns and villages such as Ramla, Lod, Kfar Qasem, and Taybe	The Bedouin reject this plan and request that either the government recognize their Syag location as agricultural villages or allow them to return to their original villages outside the Syag. A few Bedouin move to northern towns. Stalemate
Period 2: Local-urban concentrations		
1968	A plan is developed to concentrate the Negev Bedouin in seven urban new planned towns in the Negev. The neighborhoods within the towns are planned according to tribal affiliation. The first one is Tel Sheva, followed by Rahat, Segev Shalom, Arara BaNegev, Kaseifa, Hura, and Laqiya	Those who move to the towns are primarily those without land claims. The rest refuse to move
1971	The Israeli Land Authority (ILA) declares the land in the Northern Negev as state land according to a new land law passed in 1969. The ILA sets up a formal written application process whereby land claims may be filed	The Bedouin reject this decision, which they see as confiscation of their land. Approximately 3,220 land owners file land claims for 776,856 dunam

(continued)

Table 7.4 (continued)

Year	Governmental action	Bedouin reaction
1975	Governmental committee headed by Ablik defines Bedouin land in the Negev as Mawat in accordance with the Ottoman designation	The Bedouin reject this committee decision, and some go to court (e.g., the clans of El-Hawashle and Kaser Elsir). The court rules against them and they lose their land claims. Their land is designated as Mawat—state land
1980	After the 1979 Peace Treaty between Egypt and Israel, a law is passed which aims to accelerate resettlement of the Bedouin, particularly those in Tal Malaha, in order to construct military camps and airports. According to this law, the government recognizes the land rights of Bedouin in the Arara and Kaseife areas, and provides them with compensation (money and land) to leave and resettle within the two towns	The Bedouin are forced to accept this law and the evacuation; however it is accompanied with resistance and violence. The Bedouin gain some land for housing and agriculture as compensation
1986	A national public committee is appointed to explore the problem of informal building and building without permits in the Arab Sector. With regard to the Negev, the committee recommendation is to adopt the policy of concentration of the Bedouin in seven planned towns, and demolish the houses\buildings which exist outside of the town plans which at the time amounted to about 6,600 structures. In 1988 the Israeli Government accepted the committee's recommendation and began implementation which continues to present day	The Bedouin reject these recommendations. The Association for Support and Defense of Bedouin Rights in Israel initiates an alternative plan to resettle the Bedouin. This plan, prepared by Khamaisi (1990), suggests 22 different locations, expanding existing towns, including Bedouin neighborhoods proximate to Jewish towns within Jewish jurisdictions, and recognizing rural and agricultural villages, and is submitted to the Southern District Planning Committee and to the government. The plan is eventually rejected; however one of the successful purposes which the plan served was as a tool (using the government's own procedures) to delay demolition while it went through the time-consuming process of passing through the planning committees. At the same time, the Association for Support and Defense of Bedouin Rights in Israel and Khamaisi prepared a local outline plan for the unrecognized village of Dirajat which was scheduled for demolition. This was eventually accepted and Dirajat became the first village to be recognized in situ

| 1986 | The Bedouin Administration is established in 1986 as a unit of the Israeli Land Authority to deal with issues of Bedouin resettlement and land claims. Its primary purpose is negotiating compromises over land claims with the Bedouin. Over the years, various government decisions altered, expanded, or diminished the unit in terms of its role and powers. In April 1997, the Bedouin Administration was replaced with the Administration to Promote the Bedouin in the Negev: to develop settlements for the Bedouin population, promote compliance, prepare options, and promote compromise in lawsuits over land claims and as an umbrella mandate—to act to resolve the burning issues of the Bedouin population in the Negev. The change was in name only. The Israel Land Authority Council was a partner in determining the new unit's administrative roles and powers. The unit's administrative powers are reduced with time and in November 2000 the government restricts its role further. Recently the government mandates the administration to initiate plans for communities and neighborhoods in which the government intends to concentrate the Bedouin population Since its creation in 1986, the Bedouin Administration has been operating without a clear and consistent definition of roles and responsibilities. In practice, the Bedouin Administration's powers reach far beyond its designated roles and acts as the Israel Land Administration in the Negev. These roles present an inherent conflict of interest, creating tensions between the unit and the Bedouin living in the unrecognized villages—the population it is supposed to serve. It also creates tension between residents of the Bedouin towns and representatives of local authorities, because it serves as an intermediary, preventing independent access to various government agencies, especially with regard to development within the communities | The Bedouin see the Bedouin Administration as an arm of regional or local government, whose aim is to shrink the territory of the Bedouin and demolish their houses. The Israel Government grants the Administration powers to deal with all issues involving the Bedouin, including implementation of government decisions. The Bedouins view the role of the Bedouin Administration as a confiscator of their land. The Bedouin suffer at the hands of this administration and demand its abolishment. In many cases, the Administration refused to spend the budget allocated for land compensation, due to what it claimed was Bedouin noncooperation and refusal to compromise on land claims |

(continued)

Table 7.4 (continued)

Year	Governmental action	Bedouin reaction
1994	Preparation and filing of District Plan 4\14 for the Beer Sheva region. The planning principles with regard to the Bedouin sector include expanding the seven Bedouin urban towns both by way of boundary enlargement and by way of in-fill of villages located in geographic proximity, and developing new urban settlements and towns such as Hura, Laqia, and Segev Shalom. The plan is approved in 2000	The plan is rejected by Bedouin who went to the Supreme Court; however they did not succeed in either changing or stopping it. In 1997 the Bedouin established the Regional Council for the Unrecognized Villages, as a non-governmental organization (NGO) to represent the unrecognized villages. It had the support of the Bedouin local government in the seven towns and gained international support. The Supreme Court asked the government to develop a new district plan which would consider the Bedouin needs. This decision led to the preparation of a revision to the Beer Sheva District Plan—4\14\23
Period 3: Beginning of recognition of villages (some in situ) within regional councils and re-planning of the seven Bedouin towns		
2003	The government decides to recognize and develop ten villages. The process of recognition had begun and the government passes Decision 881 to establish a regional council (Abu Basma) to manage first 8, then 11, and potentially 13 newly recognized villages	Some Bedouin see this as a positive but insufficient direction. They aim for recognition of all 45 unrecognized villages along with the recognition of land rights (as reflected in the land claims) and expansion of the jurisdiction of the seven towns
2008	On 23 December 2007, the Minister of Construction and Housing, Mr. Zeev Boim, appoints a committee, headed by the former Supreme Court judge, Eliezer Goldberg, to recommend a policy for regulating Bedouin settlement in the Negev and the formulation of proposals for amendments to existing legislation. In 2008, the committee issues their recommendations, which are authorized by the government	The community is split as to whether to support or reject the recommendations but await implementation
2009	Pursuant to government decision number 1999, the Authority to Regulate the Settlement of Bedouin in the Negev is established in the Ministry of Housing. The purpose, function, and organizational structure include: Designation of new Bedouin settlements, including all aspects surrounding land claims; development of settlements, including infrastructure and public services, both for existing communities and new ones; employment integration assistance; coordination of education, welfare, and community development. This governmental authority implements governmental policy—in addition to, and often times at odds with, the Bedouin Administration	The Bedouin community continues to mistrust governmental policy and actions; they do not perceive the new authority as coming with new missions. So most Bedouin reject the authority's regulation of settlements and negotiation of land disputes. Even some overt actions resulting in the development of towns and recognized villages is not enough to gain the trust of the community. The Bedouin still view the main goal of this authority as being to resettle the unrecognized villages including evacuation of some of them

Year	Government action	Response
2011	The government appoints an Application Report (May 2011) team headed by Mr. Ehud Prawer of the Department of Policy Planning in the Prime Minister's Office, to develop and submit a detailed plan for implementation and regulation of the Bedouin in the Negev, based on recommendations by the Goldberg Committee. The purpose is to translate general principles set forth in the Goldberg report to implementable policies for settling disputes. On 11.9.11, the government approves Decision no. 3707 adopting the report of the implementation team of the Goldberg Commission recommendations regarding the issue of the Bedouin in the Negev (Naqab)—referred to as "the Prawer Plan"	The communities, particularly those in the unrecognized villages and land claimers, reject the Prawer plan as not reflecting or implementing the Goldberg recommendations. The Regional Council for the Unrecognized Villages and Bimkom (Planners for Planning Rights, an Israeli nonprofit organization) develops an alternative plan for recognizing and planning the Bedouin villages, reflecting their understanding of the Goldberg recommendations
2012	The government decides to divide the Abu Basma regional council into two regional councils: Neve Midbar and Al Kasum. This move postpones elections for heads of councils for another 4 or 9 years (by Israeli law, all new councils have appointed, not elected, heads for a term of 4 years with a possible extension of another four)	Members of the recognized villages, save the largest, oppose elections (unwilling to be governed by another hamula) and support the division of Abu Basma as a vehicle for staving off elections. Members of non-recognized villages and NGOs oppose the move to postpone democratic elections
2012	The Minister of Interior appoints a six-member Investigation Committee to Examine the Municipal and Planning Boundaries of the Bedouin in the Beer Sheva area. Their final recommendations will be submitted in January 2015	One hundred and thirty-three stakeholders appear before the committee, each presenting their perspective
2013	The government decides to convert the recommendations of the Prawer plan to law. Former Minister Zeev Begin spends a year meeting with the Bedouin community negotiating how the law is to be implemented. Begin resigns when he sees the wording of the final bill—twisting his recommendations	The community (a strange coalition of Arabs and left-wing Jews along with Jewish right-wing extremists) objects strongly to the proposed law for Bedouin resettlement
2014	The government appoints the Minister of Agricultural to implement the plans for resettlement of the Bedouin, with some amendments to the law	The land owners continue to claim their lands; the representatives of the recognized towns and villages demand more resources and expansion of their jurisdictions, and creation of economic opportunities. The unrecognized villages demand a halt to housing demolitions and either recognition or return to original lands

District Plans

The District Plan of the Beer Sheva region in Israel's Southern District followed the hierarchical logic of the national plans. The latest, Sub-District plan no. 4\14\23 arose from the need to amend its predecessor, regional plan 14/4 which ignored the needs of the Bedouin.

At the end of 1994, the District Master Plan for the Southern District, 14/4, stated that the goals for the Bedouin sector include: that the Bedouin community living outside the towns should relocate to any of the seven existing towns by way of in-fill within the towns (all towns have unpopulated areas due to undecided land claims) and expansion of current jurisdictional boundaries, reducing the population size of those living outside the towns by half. Many Bedouin and NGOs filed objections to 14/4, claiming that the plan discriminated against the Bedouin, ignoring their rights and harming their dignity and the right to preserve their unique culture and tradition. The objections were rejected, and in January 2000 the plan was approved. In the same year, a high court petition was filed by the Association for Citizen Rights and residents of the unrecognized villages against the National Council for Planning and Building (Israel's national planning board) to amend District Plan 4/14 to include the planning for additional Bedouin villages in the Negev, which would take into consideration their geographical locations and number of inhabitants and allow them to preserve their way of life.

As part of a compromise between the state and the petitioners, it was agreed that the regional master plan for part of metropolitan Beer Sheva, no. 4\14\23, would revoke the ratio goals for Bedouin town resettlement and that planners would reformulate proposals for the Bedouin population in this area, as well as for those in the entire metropolitan Beer Sheva area to "preserve and develop the shape of rural settlement as one of the solutions for Bedouin inhabitants outside the existing towns." The compromise also instructed planners to meet with representatives of the Bedouin community and together examine concrete proposals for new settlements, including rural communities.

The plan defined a hierarchy of spatial locations, residential formats, and employment alternatives with the city of Beer Sheva as the core metropolis. The metropolis would provide, in accordance with the needs of the population (in its entirety, Jews and non-Jews), radial transport networks; incentives for cooperation among communities and populations; strengthening urban and suburban communities and their quality of life through a variety of services, allowing agricultural structures in some residential areas; urban renewal; employment and industrial zones; open and recreational spaces; agricultural zones; prevention of commercial centers along the roads; and infrastructure consolidation. It was to both meet growing demands for development and preserve the unique characteristics of the region: cultural heritage and social diversity while ensuring the ability to provide and deliver infrastructure.

Given what the petitioners saw as progress with the planning process, when plan amendment 14/4/23 was deposited in 2007, they cancelled their Supreme Court petition. This subdistrict plan (for the Syag region and metropolitan Beer Sheva) was only authorized on August 8, 2012, and the petitioners argued that the slow process of planning and development reinforces the concern that the Negev Bedouin suffer blatant discrimination at the hand of government plans and processes.

Although this subdistrict plan came about as the result of a Supreme Court decision to provide new solutions for Bedouin resettlement, the Bedouin reject the planning solutions suggested by the plan and, before its ratification, submitted numerous objections which were rejected. The plan allows for the recognition of additional Bedouin villages, but does not define how many and which of the unrecognized villages are to be recognized and where (in situ or elsewhere). The plan provides some criteria for possible recognition of villages in areas zoned as mixed rural landscape.

District Plan 14/4/23 is currently the official regional plan covering the area in which most of the Bedouin reside. It both guides and regulates all spatial development, including jurisdictional reorganization and resettlement of the Bedouin localities. Although the plan provides for opportunities which could contribute to paving the way to possible solutions, the Bedouin are not considering these opportunities, instead insisting on further amendments to the plan which would designate new villages for recognition and development (Fig. 7.3).

The Goldberg Committee

With each year problems grew exponentially, and what was 'the Bedouin problem' became a national problem. During the same year that District Plan 14/4/23 was deposited, in October 2007 the government requested that the Ministry of Housing and Construction establish yet another committee to formulate recommendations for recognizing and regulating Bedouin settlement in the Negev. On December 23, 2007, the minister nominated the members of the committee who included representatives of the Ministries of Interior, Finance and Agriculture, an NGO, an expert on land issues, and the majors of two of the Bedouin Negev towns (Laqiya and Rahat); no committee member was appointed from the unrecognized villages.

The committee was headed by former Supreme Court Judge Eliezer Goldberg, and the committee referred to as the 'Goldberg Committee.' Its mandate was to "submit recommendations for an expansive, comprehensive, and realizable program that sets guidelines for Bedouin settlement arrangements in the Negev including compensation levels, alternative land allocation arrangements – and that includes recommendations for legislation as needed" (Cabinet Resolution No. 1999, 2007).

On December 11, 2008, the committee released its findings in a widely publicized report (Goldberg Commission 2008). It called for recognition of the unrecognized villages and legalization of buildings built illegally because of the impossibility of obtaining building permits. Legalizing these buildings would end the common practice of house demolitions and would also require the state to install basic infrastructures for

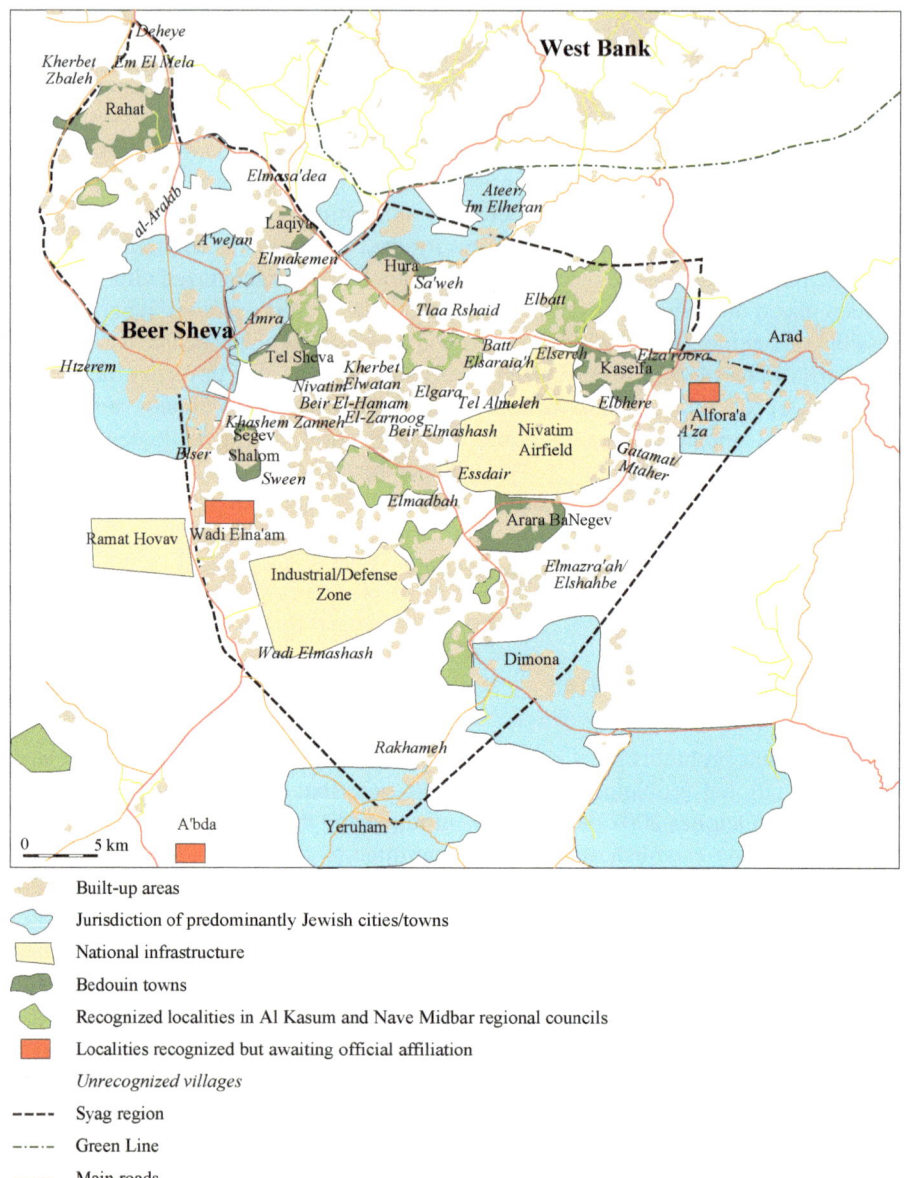

Fig. 7.3 Geographic dispersal of the unrecognized villages

them. The committee also sharply criticized the state's policies toward the Bedouin, including attempting to concentrate the entire population into the existing towns and new villages. Moreover, the report overtly acknowledged the state's continued and systematic violation of the Bedouin's basic rights. Key points include the following:

- Recognize parts of the villages according to threshold of number of inhabitants and mix/join them as possible among existing localities.
- Allow some of the residents of the unrecognized villages who request to move their village to other fixed locations in the North of the Negev to do so, and enable them to choose the type of settlement which they would like to develop.
- Define existing buildings located within the boundaries of authorized local plans as 'gray buildings,' on the condition that the buildings do not contradict plan implementation or development infrastructures. This definition would legalize these buildings and enable them to be connected to infrastructure and services, prior to obtaining the building permits which is a lengthy process.
- Produce a "quick implementation legal instrument" which would enable the recommendations of the committee to be speedily implemented.
- Establish a special planning committee, alongside the South District planning committee, to resettle the dispersed Bedouin villages.
- Recognize partially and conditionally some of the Bedouin land claims on the basis of existing usage (catchment, like squatters), and pay compensation for some of the disputed claimed lands.
- Decisive enforcement of the law and the rules regarding housing demolition for buildings built illegally during the period in which Goldberg recommendations await government authorization.

The committee did not recognize the Bedouin historical unwritten claims (indigenous) to the lands and precluded any possibility of residents taking legal steps to gain official recognition of ownership or compensation based on non-written agreements. However it did recognize historical usage rights and set forth two conditions for recognition of villages: that each village has a "minimal mass" of residents and that the village does not prevent the authorities from carrying out current national or regional plans (i.e., transportation and environmental infrastructure, army bases, urban expansion areas). The report lists the amount of compensation which will be required, including restitution of land. The commission proposes to establish an institution whose purpose would be settlement planning for the Negev Bedouin. Although the committee recognized the importance of finding solutions, the recommendations did not specify ones for a large number of unrecognized villages. Its recommendations also did not include solutions to the absence of services to residents, such as supply of electricity and water, garbage collection, and healthcare.

Public reaction was mixed (Dor-On 2009)—Soffer reflects the fears that its implementation from a strategic point of view will result in Israel's loss of control over the northern Negev (Soffer 2009), and others see this as a potential step toward solution (Zandberg 2009; Altman and Arbel 2010).

In January 2009 the government accepted the Goldberg Committee recommendations (decision no. 4441) and formed an implementation committee. Meanwhile, the number of building demolitions in both the unrecognized villages and those of Abu Basma increased dramatically. In 2010, over 700 illegal Bedouin structures were demolished, and 9,000 dunam of land were ploughed under for lack of per-mits—three times the demolition rate of each of the previous 2 years (Yagna 2010; Khoury and Yagna 2010). In addition, a group of Jewish Negev residents petitioned the Supreme Court for an injunction against the regional council head of Abu Basma for failure to demolish illegal buildings, although such buildings are part of the plans for Bedouin villages currently awaiting a lengthy planning approval process (File 2219/10 to the Supreme Court). The petition was eventually dismissed on May 26, 2013, after the Supreme Court reviewed recommendations from the Judicial Advisor for the Government.

Prawer and Begin Implementation Committees (of Goldberg Recommendations)

The Prawer implementation committee recommendations, widely seen as ignoring the principles of the Goldberg Committee they were mandated to translate, were followed by government appointment of Minister Zeev Begin to convert compo-nents of the Prawer recommendations to legislation. He spent a year in discussions with many members of the Bedouin community to draft the legislation and Begin resigned before the vote, feeling that the government had twisted the meaning of the law to an extent that the intent was unrecognizable. It passed the first round in the Israeli Knesset—parliament.

The law as submitted contradicted the main principle of the Goldberg Committee—the prompt recognition of the majority of unrecognized villages. The law defines its purpose as "to regulate the ownership of land in the Negev with respect to claims filed by owners of the Bedouin population in the Negev, in order to allow the issue of settlement of Bedouin in accordance with government deci-sions," as well as "the development of the Negev for all its inhabitants, including the periphery which can provide solutions to the Bedouin population in the Negev" (Article 1). In practice, the proposal outlines a framework for moving forward on two parallel issues:

• The unrecognized villages in the Negev, with the underlying assumption being that the residents of the unrecognized villages are invaders and all land rights belong to the state and thus the Bedouin must evacuate. The draft suggests removal of the vast majority of the villages.
• Land ownership, with principles reverting solely to state law ("the letter of the law"): The settlements suggested (monetary and in-kind land) are based on the negation of Bedouin rights to property and ties to the land. The law met with a lot of resistance with ironically traditional adversaries, the Jewish extreme right, and the Bedouin and Arab minority and Jewish left, coalescing in opposition. The Jewish extreme right rejected the law as it acknowledged some rights of the Bedouin to land and enabled the recognition of some of the unrecognized vil-lages. The Arab Knesset members and the Bedouin community leadership opposed the law for three reasons:

- The solution of the Bedouin issues and problems cannot and should not be determined by law. The law would render the Bedouin residing in unrecognized villages as criminals. Solutions need time.
- The law would overrule the Ottoman, British Mandate, and 1969 Israeli Land Law, and enable confiscation of land still in dispute.
- The law would mean evacuation of most unrecognized villages and imposed resettlement.

Investigation Committee to Examine the Municipal and Planning Boundaries of the Bedouin in the Beer Sheva Area

The issue of reshaping and demarcation of municipal jurisdictions adds other dimensions to the spatial dispute between the Bedouin and the state. On April 23, 2012, Amram Kalagi, the Director of the Ministry of the Interior, appointed a Commission of Inquiry to examine and make recommendations regarding areas of municipal jurisdiction and local planning in the Bedouin sector in the Beer Sheva district, following the Goldberg Committee principles. The Commission of Inquiry was established to promote the attainment of government objectives, as detailed in decision No. 3707 on September 11, 2011, which accelerated action to regulate the settlement of Bedouin in the Negev. The committee held about 12 meetings, hearing 133 stakeholders from the Bedouin community, government ministries and units, and a variety of NGOs. The Ministry of Interior decided to divide the Abu Basma regional council into two: Nave Midbar and Al Kasum (see Fig. 7.2). The committee continued to operate and meet with stakeholders but once that decision was taken by the government, the committee's mandate was seen as less urgent. The committee's recommendations, submitted in January 2015, begin with: "The Committee is not able to present concrete recommendations on demarcating municipal jurisdictions." It goes on to recommend the immediate establishment of local councils in Al Kasum and Nave Midbar regional councils and to provide sufficient resources for them to perform their tasks. The next stage is to hold elections for the heads of regional councils (and not delay with further appointments of council heads). There will be a need to split the two new regional councils again only after real progress is made with regard to recognition of additional villages and not as a means of delaying elections.

Alternative Planning from Grassroots Up

The responses of the Bedouin to governmental policies, plans, and actions have been more reactive than proactive as is seen in Table 7.3. However there have been some grassroots initiatives. The first was done by Khamaisi (1990). This plan suggested an alternative concept based on the planning of diverse urban and rural localities to resettle the Bedouin communities at that time. It was rejected by the government. The next major initiative occurred in 2010, when the Regional Council for the Unrecognized Villages (RCUV) and Bimkom—Planners for Planning Rights, in collaboration with Sidreh—the Bedouin Arab Women's Organization of the Negev, with support of the EU, formulated a Master Plan for the unrecognized Bedouin villages in the Negev. Unlike the government plans, the Master Plan promotes recognizing all of the unrecognized villages as possible, desirable and necessary. The departure point for the plan is that the Bedouin are not "squatters," they have lived in the region for many generations, and as citizens of the state, they are entitled to equal rights. This must include full respect for their planning and development rights, which should not be conditioned on the complex and lengthy procedure of resolving land ownership issues. The plan's underlying assumption is that a fair and agreed-upon solution, achieved in collaboration with the residents, is possible and within reach. The solution would be based on accepted professional norms, as well as on principles of recognition, equality, and justice, which would also begin to reduce the vast gaps between the Bedouin and Jewish populations in the Syag area.

The Master Plan proposes a range of feasible means for recognizing and developing all 46 Bedouin villages; of these 35 villages are still unrecognized, and comprise the heart of this plan. The population of all the Bedouin villages (recognized and unrecognized) in the plan's target year, 2030, will total some 235,000 people, out of an overall anticipated population of 440,000 Bedouin in the Beer Sheva area (including those in the Bedouin townships).

At the local level, the plan seeks to promote the Bedouin village as a distinct type of locality that will be recognized and codified by the Israeli planning system, in a manner similar to the Jewish 'moshav,' 'kibbutz,' or 'yeshuv kihilati' rural typology. Such recognition will integrate both the historical and the planning logic of the Bedouin village. To this end, the plan outlines, for the first time, a model for the development of the Bedouin village. This model is based on (1) attachment forged between communities and their living space; (2) the system of traditional land inheritance in which the land is divided between tribes, clans, and extended families; and (3) the location and function of the open spaces, roads, and public institutions within these villages.

This alternative plan illustrates how the villages can be developed according to density and zoning criteria, as is accepted in Israeli Governmental plans, with adjustments according to Bedouin spatial norms. The result is planning that employs, as a starting point, the existing development patterns, and yet seeks to densify development in the future. This is needed in order to meet the minimum population size required for the provision of public services. The model addresses the possibility of extending the village neighborhoods, as well as of providing services to distant clusters of families. The plan also outlines the desired development structure of roads, public institutions, and open spaces in the villages (Yiftachel et al. 2012, updated 2014).

This alternative regional plan has been submitted to government agencies as a basis for negotiation which would lead to voluntary resettlement of the unrecognized villages in the Negev. To date the official governmental planning system has ignored this counterplan, while the Bedouin community claims to accept the principles and guidelines it purports.

Chapter 8
Lessons Learned

Understanding the historical contexts of government resettlement policies and Bedouin reactions is what guides suggestions for the alternative policies and actions which will be presented in the concluding chapter. Policies which take into consideration territorial, functional, and sociocultural dimensions of the issues reviewed in this volume may well reduce tensions between the state and Bedouin communities, and facilitate resettlement, meeting interests of both sides.

All policies to date—regional concentration in the Syag in the 1950s as well as encouraging Bedouin outmigration to the northern and central areas of Israel or to the West Bank and from there to Jordan (Shmueli 1980); urbanization in seven new towns beginning in the late 1960s; recognition of villages—in situ and relocation— as part of regional councils beginning in 2003; and the recent expansion of Bedouin town jurisdictions still leave 3,220 unresolved land claims for 776,856 dunam and close to 56,000 Bedouin living in unrecognized villages in a sprawling jumble of tin and cinder-block shacks, livestock pens, lean-tos, and refuse dumps with no rights to municipal services, such as running water, electricity, sewage, or garbage collection. Widespread poverty and social neglect as well as rampant crime are the reality in both recognized towns and villages, and in unrecognized Negev Bedouin communities. These policies have all been developed in a top-down manner with limited or no participation of, or collaboration with, the Bedouin communities themselves.

After decades of neglect, governmental agencies are aware of the pressing need for effective planning and development policies which include the management, if not resolution, of land disputes within the communities. The policies to date ignored many of the Bedouin needs and indeed the communities were 'invisible' in the overall regional plans for the Negev and the metropolitan Beer Sheva area. This has resulted in alienation of the Bedouin and their reactions to governmental spatial policies with mistrust, fear, and rejection. Between 2009 and 2013, resistance to the Prawer plan and the Bedouin Resettlement Law (2013) are two such recent examples of policies strongly opposed by Bedouin and their advocates. The first example of counterplanning is traced back to 1990, when the Association for Support and

© Springer International Publishing Switzerland 2015   69
D.F. Shmueli, R. Khamaisi, *Israel's Invisible Negev Bedouin*,
SpringerBriefs in Geography, DOI 10.1007/978-3-319-16820-3_8

Defense of Bedouin Rights in Israel prepared an alternative to the government master plan for resettlement of the Bedouin localities (Khamaisi 1990), and later in 2012, when the Regional Council for the Unrecognized Villages and Bimkom prepared a comprehensive alternative plan to recognize the remainder of unrecognized villages in the Negev (Yiftachel et al. 2012, updated 2014).

The series of forced resettlement plans with limited options, ignoring customary law, socio-spatial structure, and traditional cultural and economic functions, has led to a deepening of the conflict between the state and its citizens. Policies and plans developed with the Bedouin communities, with varied options which meet the growing diversity within the community, are crucial—alternatives which preserve current social structures and clan affiliations as well as options which develop opportunities for emerging new community structures and hierarchies to meet the needs of a younger educated generation and a middle class where individualism, nuclear family, and an expanded range of professions are gaining prominence. This chapter lays the framework for this diversity of planning alternatives by learning from what has happened to date.

The government's policy of Mawat (using the Ottoman land regime definition of the Negev as 'dead land'), converting it to state land and Judaizing the Negev has been the overriding territorial and demographic ideology guiding all plans and policies (Yiftachel et al. 2012, updated 2014). Mawat has been used by both the government and the judiciary to confiscate and shrink what the Bedouin consider their territorial space by customary law, for the benefit of new Jewish towns and villages and national projects—industrial zones, military and airport bases, and infrastructure. National government ratification of the Goldberg Committee recommendations of 2008 marks a clear recognition of unfairness and injustices of past policies. This realization has occurred among some government agencies but not all. Particularly on the district level, some bureaucrats have not internalized this transformation and discourse continues to relate to the Bedouin as strangers who build without permits on state land. Government has been painfully slow in recognizing more villages according to the Goldberg criteria, facilitating development of recognized villages and investing in their infrastructure, and removing barriers to economic development. Some Bedouin have also not internalized the transformation and continue to react as opposed to initiate solutions. The theoretical framework set forth in Chap. 4 will be used to structure the analysis of lessons learned.

Land Claims of Indigenous People

The Negev Bedouin define themselves as an indigenous people. Sanders in 1999 looked at the changing situation of the Negev Bedouin and their indigenous status. Recently there have been academic arguments with Yahel et al. (2013) arguing that the Bedouin are not indigenous to the Negev and Amara et al. (2012) bringing counterarguments in support of their indigenous status. Regardless of the academic arguments, some of the principles set forth in Chap. 4 are applicable to the Negev Bedouin. They are undergoing rapid selective urbanization and modernization and run the gamut between preservation of traditions and opening up to the Westernization and modernity surrounding them.

The Bedouin are using the recent focus on the indigenous affiliation as a tool for securing land claims and for preserving cultural and tribal relationships to obtain their territory and survival as a sociocultural group in the face of aggressive governmental policy and action to fragment the community and clan ties. Today the definition of indigenous may help secure the aspirations of some of the Bedouin tribes still living in small unrecognized villages, but isn't applicable to many of the Bedouin living in towns and large villages. For those in the unrecognized villages seeking recognition, many see land as fundamental to their survival both culturally and economically.

While we support many of the Bedouin land claims and the aspirations to preserve the sociocultural fabric, they must go hand in hand with options to enable development and open new opportunities for the majority of the Bedouin in the towns and regional councils and for the educated younger generations. The current rhetoric within the community does not yet reflect this need but remains focused on land claims and territorial needs and aspirations.

There currently exists a fragmented Bedouin leadership—not a unified one speaking for the diversity of needs which could negotiate with government agencies against the backdrop of widely agreed-upon principles for the expanding community (Shmueli et al. 2009). Thus government agencies approach individuals, or individuals appear before committees lobbying for their very particular requests/needs. There is no holistic strategy from which principles have been developed. This enables the government negotiators to employ a divide-and-conquer methodology by addressing land claims on an individual basis—offering different amounts for similar parcels of land, in return for abrogating land claims. The government's strategy with regard to land is primarily to ignore the Bedouin's claims to ownership except when an individual is amenable to selling his claim to the government.

The conflict over land ownership is presented by both the government and leading Bedouin representatives as the main component of the dispute between the state and its Bedouin citizens. This is only part of the truth. Other key components include economic hardship, diversified needs within the community, strong connection made by the government between resolution of land ownership issues as a precondition for planning-zoning and enabling development (by issuing building permits), and demarcation of municipal jurisdictions. There is also a shifting geopolitical and national affiliation—from strong association with the Jewish majority and the state to a closer identification with the non-Bedouin Arab minority.

Urbanism, Modernization, Westernization, and Forced Urbanization

The seven Bedouin towns were planned and populated as 'tribal towns,' with clans belonging to the same main tribe residing in one town, with homogeneity of clans and extended families at the town neighborhood level, preserving the social structure of tribal affiliation in a smaller geographical area. Thus for instance, Segev Shalom is populated mainly by the Azazmah tribe, Rahat by the Tarabin, and Kaseifa by the Tiaha. This policy, thought to be culturally sensitive in that it was attractive to the Bedouin willing to move only as a group, has led to false urbanization along two dimensions: social behavior and structural stagnation, and lack of an economic municipal base. All of the towns are unofficially 'closed' to other tribes and the largest clan controls the town. The policy, successful in gathering the tribes originally, today contributes to social segregation between and within towns, and immobility—stymied immigration from rural areas to towns. None of the towns have the economic base to support their inhabitants and are ranked among the poorest in Israel. Rahat, for instance, the largest of the Bedouin towns with approximately 60,000 inhabitants, is still divided into clan neighborhoods with no geographic mobility amongst the neighborhoods. The population of the neighborhoods are continually growing without additional land area, and suffer from an insufficient and weak housing stock and economic hardship. There is little urbanism, and familiarity with other, more 'Western,' cities in the country only augments the failures of the Negev Bedouin towns. Those Bedouin in the unrecognized villages see the towns as a failed option and wait for other alternatives—different urban models and rural villages.

Justice: Distribution, Recognition, Participation, Compensation

Justice is analyzed by looking at four aspects: distributive issues as they relate to land allocation and use, recognition, participation, and compensation. The lessons accrued combine the need for land management based on functional need, recognition of the land claims entailing either granting of official ownership or other solutions including payment of fair compensation-monetary and in-kind – in cases where the government decides that the specific land is needed for other national purposes, and all this done in collaboration with the Bedouin community.

    The criteria for recognition must combine an understanding of customary Bedouin law which encompasses elements of Sharia' law, and state Law. Recognition by the government of Bedouin land rights (ownership or usage) would enable both sides to develop criteria for just compensation in lieu of granting all land claims. This can only be done collaboratively with Bedouin participation in developing criteria. The Bedouin, as a peripheral and marginal traditional tribal society, cannot hope to reduce the economic and social gaps between Bedouin and Jewish localities without large government investment in collectives (towns, clans, tribes) and significant aid to individuals when it comes to development of infrastructure and building of permanent homes. The compensation to individuals should be based on the role of the central government and its district agencies in providing public services and infrastructure for the Bedouin. Without intensive and special governmental support to close the gaps and facilitate future development, the Bedouin will not be able to break the sociocultural, structural, ethno-national, and geopolitical barriers (Khamaisi 2013).

Modular and diverse approaches are called for, including:

- Recognition in situ (*distribution*).
- Reallocation of land resources by first recognition of the rights of those holding Ottoman land titles, and second by a combination of customary and state laws as a basis for settling the remaining land disputes (*distribution and recognition*). Such settlement packages might include (*compensation*):

  - Just compensation which considers land betterments by zoning—allocation of alternative land for purposes of housing and for an economic base. When the government evacuated Bedouin from Tal Malhat (in order to build the Nevatim Airport Base) to Arara BaNegev and Kaseifa in 1980, the evacuees were given compensation in the form of alternative lands zoned so as to enable development of agricultural economic activities. Land in other locations, zoned in a way to promote development (of diverse types), is one form of such compensation.
  - Government payment of fair monitory value for land that it decides cannot be returned to the landowners based on the land's potential, not current, value.
  - Compensation might include additional land zoned for development to incorporate growth of future generations.

- Re-demarcation of current Bedouin jurisdictions for multiple purposes (*distribution*):

  - To increase potential revenue and allow the development of employment zones.
  - To accommodate the existing population.
  - To include some of the unrecognized villages as neighborhoods.
  - Boundaries which allow for the planning, development, and management of joint (Bedouin-Bedouin; Bedouin-Jewish) industrial zones to secure reallocation of the public revenue on equal and equitable bases.

- The recognition and development of new villages, currently invisible in regional plans, with appropriate planning, development, and investment (*development and recognition*).

Distributive Issues

In terms of *distribution*, the resounding message is the need for a varied set of options which fit the diverse needs of the Negev Bedouin community and enables their transformation from a regional burden to a regional benefit. Government policies have historically aimed at rapid urbanization of the Bedouin community, initially offering only one type of solution. That solution did include consideration of social and tribal affiliation, but it ignored the diversity among the Bedouin, which requires a variety of municipal options, some freedom for the individual or part of a clan to choose the type of locality (different urban and rural alternatives) in which they want to live. Most of the Bedouin communities which remain today in unrecognized villages require a rural lifestyle and small municipal framework. Initiating new plans for unrecognized villages should be undertaken separately from land claims, particularly since a significant number of the Bedouin families do not own additional lands for development.

Recognition

The main conflict with the Bedouin surrounds the land ownership disputes. The Israeli Government does not *recognize* Bedouin land rights over land, nor does it recognize principles based in customary and habitual Bedouin law. It does not recognize the Bedouin as a traditional or indigenous community whose culture should be preserved.

The common discourse among Israeli planners and policy makers is of the Bedouin as a burden and barrier to Negev development. Ariel Sharon, as Prime Minister is quoted (Abu Ras 2006) as saying:

> In the South we face a very difficult problem: approximately 900 thousand dunam of land in the country is not ours but controlled by the Bedouin population. I, as a resident of the Negev, see this problem every day. This is actually a demographic phenomenon and… Parties may not be alert to this issue. We, as a country, do nothing against this phenomenon. The Bedouin occupy new areas, reside in the reserve lands of the State – and no one is doing anything significant about it. (Ariel Sharon, in Abu Ras 2006).

Dealing with the Bedouin as a burden and problem, excluded from public space and the regional economy, has a direct and negative impact on the Bedouin, particularly the younger generation. The government must recognize the Bedouin needs and see them as potential contributors to the region's economic development, moving from burden to leverage. Policies of equity and equality must be put in place to make this change not symbolic but functional, contributing to a change in behavior and reduction or even end to the culture of poverty which pervades the Negev Bedouin community.

Participation

These types of changes would require *participation* of the local population, actually Bedouin collaboration with government planners to develop a culturally oriented set of alternatives.

Compensation

Settling of land disputes could be achieved with equitable *compensation* based on customary as well as state law which might include alternative lands and monetary compensation as outlined above.

Activism in the form of initiatives, public pressure by the community, and non-governmental actors, combined with use of the courts to demand justice, are what have led the government to increased responsiveness and recognition of Bedouin needs in recent years. There is still a ways to go.

Chapter 9
Proposals for Flexible Bedouin Resettlement and Collaborative Planning

Many transformations are emerging within the Bedouin community which should guide the formulation of future policies. The Bedouin today are a much more diverse community and Bedouin stakeholders and interests can no longer be classified and understood solely according to traditional tribal/clan structure and behaviors (Abu-Rabia-Queder 2006; Abu-Saad et al. 2007; Abu-Rabia 2012). Given this greater diversity and lack of trust among all involved, collaborative planning and development are imperative—not the top-down policies and planning processes which have been unsuccessful to date. The following is a summary of transformations and their implications:

• Population growth: Although the rate has been decreasing slowly, Bedouin population growth remains quite high when compared to other population groups within Israel. The Bedouin population will double in the next two decades and resettlement policy and municipal structures must plan for this reality. This growth will increase the diversity among the Bedouin, which implies the need for a variety of rural and urban settlement options, including communities of various sizes and in different locations. Clan/tribal affiliations will shift and modern demands for individual (as opposed to clan) needs will increase. Currently small recognized villages will grow into municipal self-governing ones, and unrecognized villages will meet the demographic criteria for recognition (Khamaisi and Shmueli 2009).

Implications: Both governmental and alternative community spatial planning policies must recognize these population dynamics and not consider the villages as static entities. The population growth can either become a leverage for development or a burden on both the population and the region. Resettlement policies which do not consider population growth and diversity will not meet the needs of the Bedouin communities in particular and those of the Negev region in general.

© Springer International Publishing Switzerland 2015
D.F. Shmueli, R. Khamaisi, *Israel's Invisible Negev Bedouin*,
SpringerBriefs in Geography, DOI 10.1007/978-3-319-16820-3_9

- Land claims and planning processes: Regardless of the number of land claims which will eventually be recognized by the government, the litigation process will go on for years. Within Bedouin society and particularly among the land owners or claimers, land inheritance is acknowledged according to Islamic (Sharia') law leading to fragmentation of the land. Successors could lose the right to claim the land according to Israeli State law. Beginning with the grandparents who claim the land, the next generations may not be able to provide evidence of ownership according to state law or court requirements. At the same time, the numbers of inheritors will grow significantly with each generation, the inheritors evolving with time from nuclear into extended families and clans. In such situations, the motivation of individuals to claim small pieces of land or parcels will shrink because the parcel size is too small to make a claim worthwhile. This social structure, legal, and land disputes create significant barriers to resolution and demand time to resolve.

Implications: Planning and development processes should be separated from the legal process, and a way to provide building licenses for permanent structures and public infrastructure and space which comply with approved outline plans must be established. To date, government policy has been to concentrate first on the legal issues surrounding land claims, and only later to address planning issues. We suggest another approach which addresses planning and legal issues in parallel—while attempting to settle land ownership disputes (including those among the Bedouin themselves resulting from inheritance issues) the government, together with the community, should prepare plans, recognize new villages, establish local councils, or annex newly recognized villages to existing local, regional, or municipal entities (Bedouin or Jewish), and/or allocated jurisdiction area for managing development, and providing building permits.

- Increase of non-land-owning Bedouin: Ninety-three percent of Israel's lands nationwide are leased for 99 years. Given Bedouin demographics, the number of Bedouin who do not own land will grow considerably and thus an understanding of these groups' interests is paramount as young households are in urgent need of housing solutions. This new landless generation supports the land claims of the clans, but there is a growing awareness among them of the difficulties. Settling land disputes within the context of the ethno-national conflict and the complexities and problems inherent in the interface between state, customary and religious (Sharia') laws (Fig. 9.1) is an extensive and complicated process accompanied by tensions and conflicting expectations. An increasing number of Bedouin have no land claims and are landless.

Implication: For these people the state needs to allocate state land. If the landless group is currently living on state land which is not claimed by another clan or individual or slated for national infrastructure, the community often feels an attachment to that place and, in the case of unrecognized villages, wants recognition in situ and the accompanying rights to housing and services.

Fig. 9.1 Three legal systems
which apply to land disputes

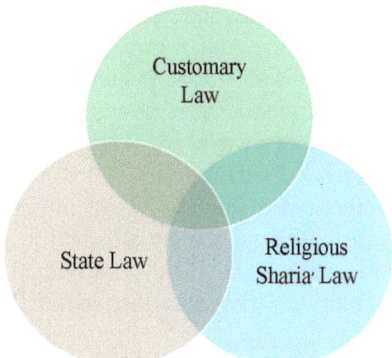

- Tribal structure is still strong, but new social structures are emerging based on geographic and socioeconomic status, locality, or neighborhood affiliations. Despite discrimination and resulting barriers, the Bedouin have begun to develop a civic society. In the lead are a number of national and local NGOs, particularly in the planned towns of Hura and Rahat. Two-thirds of the Negev Bedouin live in either the seven towns or the recognized villages, and, having made significant residential changes, desire regional functional and socioeconomic integration. The transformation has been from a uniform tribal society to a diverse community in terms of rural/urban lifestyles, education, employment, and social status. This is similar to the transformation undergone by Arab communities in Israel in general, and the Bedouin of the Galilee since the 1980s. Among the Bedouin of the Galilee the strength of tribal affiliation has decreased as the importance of municipal association and development has increased (Sternberg Ben-Baruch 2013).

Implications: Planning and development opportunities among the Bedouin in the Negev have to facilitate this type of natural transformation, and prepare planning on the local and district levels for what will follow traditional structures and behaviors. There is a need to deal with the Bedouin issues differently than in the past, solve present problems, and prepare for the future.

- Economic and employment structure: Many of the Negev Bedouin today still work in agriculture. The Bedouin suffer from high unemployment, low incomes, and high poverty levels and are part of a marginal regional economy. Despite this, they are undergoing rapid economic transformation, benefitting from the economic opportunities resulting from the government policy of giving high priority to the Negev region's economic development.

Implications: Planning should encourage potential economic opportunities, in addition to agriculture.

- Modernization with or without urbanization: The official government resettlement policy still focuses on creating accelerated urbanization among the Bedouin. This policy began in the late 1960s with the planning and development of the seven towns, and changed somewhat with the recognition of semi-urban localities within the Abu Basma regional council in 2003 (divided into the two regional councils of Al Kasum and Neve Midbar in 2012). One of the goals of urbanization was to modernize the traditional community. The community responded in contradictory ways, some accepting and others refusing to move. The explicit catalysts behind the policy of urbanization were functional and territorial. Additionally, there were implicit considerations such as demographic and social change—from a traditional society attached to place according to tribal structure to a modern society based on the individual who is both geographically and functionally mobile. Urban localities require less area in which to house more population.

The Negev Bedouin urban resettlement policies systematically followed the spatial and demographic principles developed by the government. They were shaped from top-down, ignoring community participation, assuming that urbanization would lead to modernization in a short period, enable the government to provide municipal services at a low cost, and develop a new system of relationships among individuals, including reduction of population growth. An illustrative quote is from Moshe Dayan, at the time Minister of Agriculture:

> We should convert the Bedouin to urban workers— in industry, services, construction and agriculture. Eighty-eight percent of residents of the State of Israel are not farmers. The Bedouin should be part of this (non-agricultural) majority. Indeed this will be dramatic change. It is means that the Bedouin would not live on his land or with his animals. He will return home in the afternoon and put on slippers. His children will become accustomed to their father wearing trousers, without a knife and not removing lice from his hair in public. They will go to a school with combed hair and a part. This will be a revolution, but can be done in two generations. Not by force or coercion but with governmental direction. This reality would mean that Bedouizem would disappear.
>
>   Moshe Dayan, former Minister of Agriculture, HaAretz, July 31, 1963 (translation from Hebrew).

This policy of enforced urbanization to achieve territorial goals in fact occurred with limited urbanism. In parallel, however, the community began to adopt a selective type of urbanization and modernization which deviates from the Western model of urbanization (see Fig. 4.1, Chap. 4). The development of the Bedouin community, as with other communities in similar circumstances, ranges from rural to urban, villages to towns, and traditional to selective modernity.

Implications: Resettlement policies should facilitate diversity along this spectrum, and weaken the relationship between urbanization and modernity. Modernity can occur in rural communities, particularly in view of the increasing global economy and the short distances between localities in the metropolitan Beer Sheva area. Most of the Bedouin communities are located in the middle and outer belt of this rapidly developing metropolitan Beer Sheva area.

- From reaction to initiation—strengthening of civic society: The situation today is quite different from that of 1948 with the resettlement of the Bedouin tribes in the Syag area. Some clans received oral or written commitments at the time that they would be able to return to their lands and villages once the new nation of Israel was stabilized. These commitments were never met. All clans were resettled. In the 1960s, the reaction of the Bedouin to the urbanization policy differed according to tribal affiliation, land claims, and social status. Some moved, and others refused. Today however, the Bedouin community is no longer only reacting to government policy; it is proactive and is initiating plans. This was made possible by changes within the community resulting from population growth, increasing levels of education—including among women, socioeconomic change, growth of personal income, and some normative changes among young communities. Outside factors contributing to the strengthening of Bedouin civic society include the democratic system in Israel as whole and the involvement of national NGOs in what is happening in the Negev, particularly toward the Bedouin communities. Media coverage and the growing international support for securing humanitarian needs of marginal communities contribute to the change in discourse, the language and actions of advocacy, and the development of alternatives to governmental plans for Bedouin resettlement (Khamaisi 1990; Yiftachel et al. 2012, updated 2014). This new activism which includes the involvement of the Bedouin municipal leadership, national political parties, and non-governmental organizations is epitomized by the effectiveness of a coalition which opposed the Prawer plan between 2011 and 2013.

Implication: Policies of resettlement, spatial planning and zoning, and municipal restructuring must work with the various segments of the community to shape appropriate solutions and arrangements.

- Growing demands for equal and equitable goods and services to be provided by the local, district, and national government levels: Today the Bedouin are not satisfied with minimal municipal services which exist in the seven towns and some of the recognized villages and are nonexistent in the unrecognized villages. The young generation demands a reversal of the unjust allocation of land resources as a result of unfair zoning, limited municipal jurisdictions for Bedouin, and an unbalanced policy of lands for Jewish councils at the expense of Bedouin jurisdictions.

Implications: In shaping future resettlement, rezoning, and reorganizing the allocation of municipal resources, residents of both recognized and unrecognized villages demand an equitable basket of resources equal to those of similar Jewish jurisdictions. This reallocation must take into consideration three main components and their interface: land ownership, zoning, and land management (Fig. 9.2).

Fig. 9.2 Three main components of land allocation

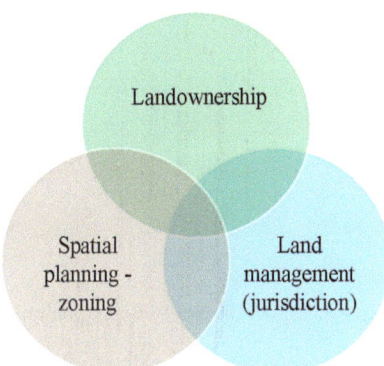

The above transformations and their implications cover aspects of all three elements. What follows are recommendations which focus on the interface of spatial planning, management, and ownership.

The Proposed Planning Model for Possible Resettlement

The planning model, put forth in 2011 (Shmueli and Khamaisi 2011) (Fig. 9.3), attempts to simplify the very complex conflicts of interests, needs, and tensions between the Bedouin (struggling for land resources, territory, and equal economic opportunities) and the national and local Israeli Governments. These governments are confronted by what they see as a growing demographic threat in the Negev, the need to interdict the smuggling of drugs and arms, and intercept terrorists and illegal immigrants using Sinai and Negev desert routes to cross into the heart of Israel. The irony is that governmental discriminatory policies are undermining the loyalty that the Bedouin had long displayed when serving as Israeli army scouts and trackers. Conflicts within and among the stakeholder groups themselves compound the difficulties of arriving at acceptable solutions.

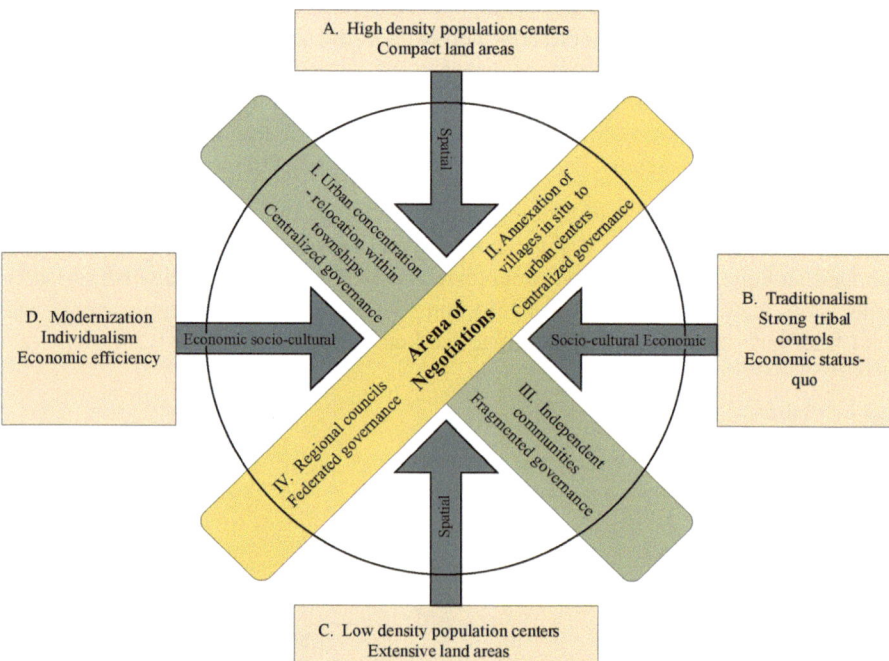

Fig. 9.3 Model for Negev Bedouin municipal planning structures

The model's vertical binodal axis is spatial, with planning options ranging from high-density population centers with compact land areas that limit the numbers of localities and the amount of land allocated for urban development (A) to low-density population centers with extensive land areas that maximize the number of localities based upon tribal and clan affiliation with large areas for development (C). The binodal horizontal axis presents the sociocultural and economic conditions that influence the planning options. It reflects the clashes between traditionalism, clinging to strong tribal controls and maintaining a very weak economic status quo (B), and modernization trends which lead to economic efficiency and infrastructure development, putting greater emphasis on family and individualism rather than tribal hierarchy (D).

The quadrants are produced by the convergence of the dominant forces that shape each half of the two axes. What emerges are four municipal forms and three governance structures. Public attention and debate continue to focus on the impasse between planning options in Quadrants I (urban concentration) and III (dispersed independent communities), while the government and Bedouin leadership are moving slowly and hesitantly toward Quadrants II (urban annexation of villages in situ) and IV (regional councils embracing dispersed settlements) which are in the early stages of implementation or under consideration. Focusing the public planning discourse on Quadrants II and IV would put greater pressure on the government to support these options more wholeheartedly, thereby enhancing the prospects for negotiated agreement.

The four quadrants reflect very different approaches to municipal structural planning.

Quadrant I (Fig. 9.4): Urban concentration in higher density population centers that create compact land areas dominated by government policy from the mid-1960s through the mid-1980s: Bedouin tribes were relocated within the seven new Bedouin towns. Related tribes/clans in nearby areas were relocated to these towns. Other clans, usually those with land claims where they had been settled, and clans antagonistic to those being relocated, were left in place. Planning for each town was based on neighborhoods populated by individual clans and extended families in a plot-based (often a dunum) settlement pattern. The town centers are usually located centrally outside the clan neighborhood and include the market, municipal buildings, health clinics, schools, and mosques. Although subcenters were not planned, over time most neighborhoods developed commercial roadside strips. Public transportation in the form of busses began to operate only this past year on the main roads of Rahat, the largest of the towns. The other towns rely on private transportation although many have roads wide enough for bus access.

Although these new urban centers reflected an attempt to maintain tribal communalism while ensuring territorial concentration, the sociocultural changes among the Bedouin, including outside work opportunities, promoted greater individualism. In each town there are islands of what the government views as available empty space. However the Bedouin reject their use since, according to Bedouin law, these spaces are owned by other tribes. The Bedouin view the towns as failures because

Fig. 9.4 Quadrant I

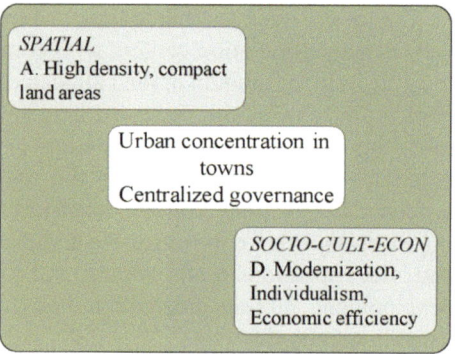

of congestion, internal tribal factionalism, limited availability of open space, and lack of industrial, recreational, or other service sites which undermine economic development possibilities. Although they were designed to create economic opportunities and cut the costs of services, they have become Israel's poorest communities because their inhabitants, while unprepared for accelerated urbanization, lost their agricultural and grazing lifestyles.

Quadrant II (Fig. 9.5): Annexations of villages in situ to urban centers would allow small communities located in close proximity to Bedouin or Jewish urban centers to be annexed administratively to them as neighborhoods. For the most part this involves leaving in place recognized and unrecognized villages and enlarging municipal jurisdictions based upon territorial contiguity.

In this quadrant the annexed villages would be planned as 'closed' neighborhoods, maintaining the identity of the neighborhood as a sub-entity and, depending on the desires of the population, either take into account the organic development already on the ground or develop their own modern spatial design. Initially the peripheral areas can be zoned for grazing and farming; zoning can be changed in the future to accommodate urbanization.

Quadrant III (Fig. 9.6): Dispersed independent communities reflect current Bedouin planning desires: recognition of unrecognized villages in situ and maintaining fragmented tribal governance. This precludes economies of scale while providing Bedouin territorial control and nurturing tradition, but offering major challenges to infrastructure and economic development. The Bedouin would like the government to recognize the existing villages, taking into consideration land ownership and/or catchment rights. Each kinship tribe prefers to live in it own social spatial area. The plans would mostly retain the existing organic structure, with the Sheikh's home more or less in the center and with large areas for grazing and farming. There is no desire for shared centers which means that each village would have a small, limited center, infrastructure would be very expensive, and some of these communities would be quite far from outside work centers.

Fig. 9.5 Quadrant II

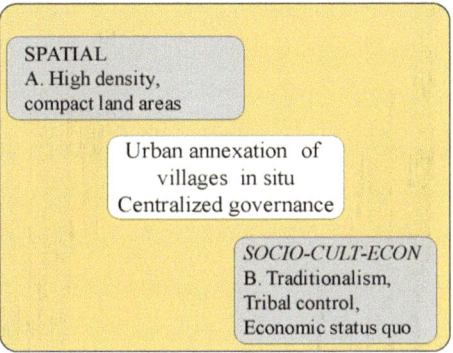

Fig. 9.6 Quadrant III

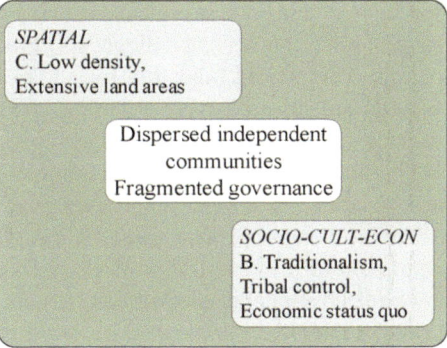

Quadrant IV (Fig. 9.7): Regional councils throughout Israel embrace forms of federated governance that are appropriate for strong small- to mid-size communities on the periphery of metropolitan centers. Abu Basma was promoted as such a decentralized regional governance structure, on the assumption that the Bedouin would wish to live within the metropolitan periphery while enjoying a rural and communal quality of life.

The model was developed and used to facilitate stakeholder debate on the impact of changes that may result from various factors:

- Population growth
- Social disintegration due to modernization and urbanization
- Processes of geographical, social, and functional mobility
- Availability of local and regional economic resources
- Presence or emergence of local leadership
- Functional organization of bodies, corporations, and unions for the management of issues that have land use and jurisdictional boundary dimensions (e.g., environment, sewage)
- Identifying competitive advantages permitting economic growth

Fig. 9.7 Quadrant IV

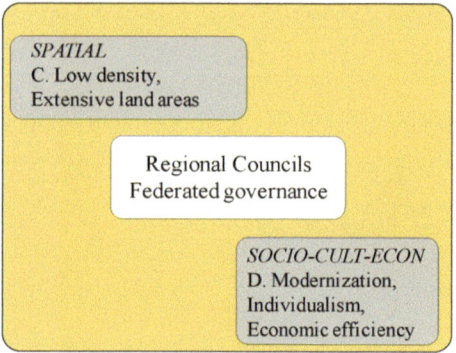

Whereas in Quadrant III there is no municipal or functional cooperation, Quadrant IV has both. A regional center is planned for the existing regional councils with commercial areas, a college, sport facility, regional council buildings and high schools. This would hold true for additional future regional councils as well. The councils are charged with planning, providing municipal services, creating joint services, developing economy, and representing individual communities with outside bodies. Each village plan includes an elementary school, a mosque, small commercial centers and mixed land uses.

These changes expose some of the planning dilemmas of the Negev Bedouin population as a whole. Supporters of Quadrants I and III are vocal and politically powerful. Proponents of Quadrant I include the Israel Land Authority (and its offshoots in the Negev), the security establishment, branches of the Interior Ministry, and leaders and residents of most of the adjoining Jewish communities and some of the Bedouin townships. Proponents of Quadrant III are traditional Bedouin tribal leadership and the Regional Council for the Unrecognized Villages of the Negev. Their position is the opposite of that of proponents of Quadrant I, which puts negotiations at an impasse.

Those supporting Quadrant II include some of the Bedouin and Jewish towns which would benefit from extending their territorial jurisdiction. Sponsors of Quadrant IV include branches of the Interior Ministry, the Neve Midbar and Al Kasum leadership, and the recognized Bedouin villages which agreed to affiliate within the councils. Some of the unauthorized settlements also favor joining the councils.

Options Supported by Field Research

Options in all but Quadrant I require expansion of areas of jurisdiction. Quadrant III would require the greatest land expansion, while in II and IV the extent of the expansion depends upon the population size, land ownership, social affiliation, and need.

Quadrants II and IV seem to hold the best prospects for negotiating planning and municipal structures. They embrace the following principles which emerged from the stakeholder interviews and workshops, and meet some of the interests of most of the stakeholders:

- Creating diverse forms of settlement to promote mobility
- Maintaining communalism and tribalism and using them for social reinforcement and economic leverage
- Extending village boundaries to include contiguous settlements as neighborhood subdivisions
- Developing functional cooperation and establishing agency frameworks that transcend municipal boundaries
- Building on existing procedures to encourage regulatory changes that will assure more equitable allocation of planning and land resources among diverse populations
- Striking a territorial balance between urbanization structures and regionalization

In Quadrant II, attaching villages to existing towns strikes a balance between governance centralization and tribal affiliation that can maintain communalism through neighborhood frameworks. For example, the Bedouin town of Rahat, with an estimated population of 60,000 inhabitants and land area of about 22,000 dunam, can be extended to include, as neighborhoods, nearby unrecognized villages of Deheye, Em El Mela, and Kherbet Zbaleh whose tribal affiliations are acceptable to Rahat's residents[1] (Fig. 7.3).

The unrecognized Bedouin village of Rakhameh with a population of 1,500 is partially located within the boundaries of the Jewish development town of Yeruham (population of 8,800) (CBS, 2014:158). In 2005 the then head of Yeruham's local council proposed that Rakhameh become a recognized neighborhood within the town and that Bedouin residents in Rakhameh's east move to western Rakhameh that lies within Yeruham's boundaries. Both the Bedouin and the town supported the idea; however the plan was dismissed by the Interior Ministry (Dayan 2010). There is still support for the annexation within Rakhameh and Yeruham (Fig. 7.3).

In Quadrant IV, the two regional councils, Neve Midbar and Al Kasum, have an estimated registrant population of 6,900 and 8,200, respectively (CBS 2014:155)— still very large for regional councils in Israel. The plan calls for their expansion by including each unauthorized settlement as it becomes recognized. As these population numbers are unwieldy, the long-term plan might be to create a total of four regional councils (including the two existing councils) (Fig. 9.8). The regional boundaries would be defined by village contiguity, orientation to nearby larger cities, and barriers to territorial expansion, including a vast defense installation separating the southwestern villages from the settlements in the center of the Syag and an airfield separating those in the central area from village clusters in the north-

---

[1] This would increase Rahat's population by 3,500 (projected to increase to 7,300 by 2030) and extend the town's land area from about 22,000 to 30,000 dunam. Similarly, Elmasa'dea, A'wejan, and Elmakemen, with estimated populations of 4,100 (projected to 8,500 in 2030) can be annexed to Laqiya.

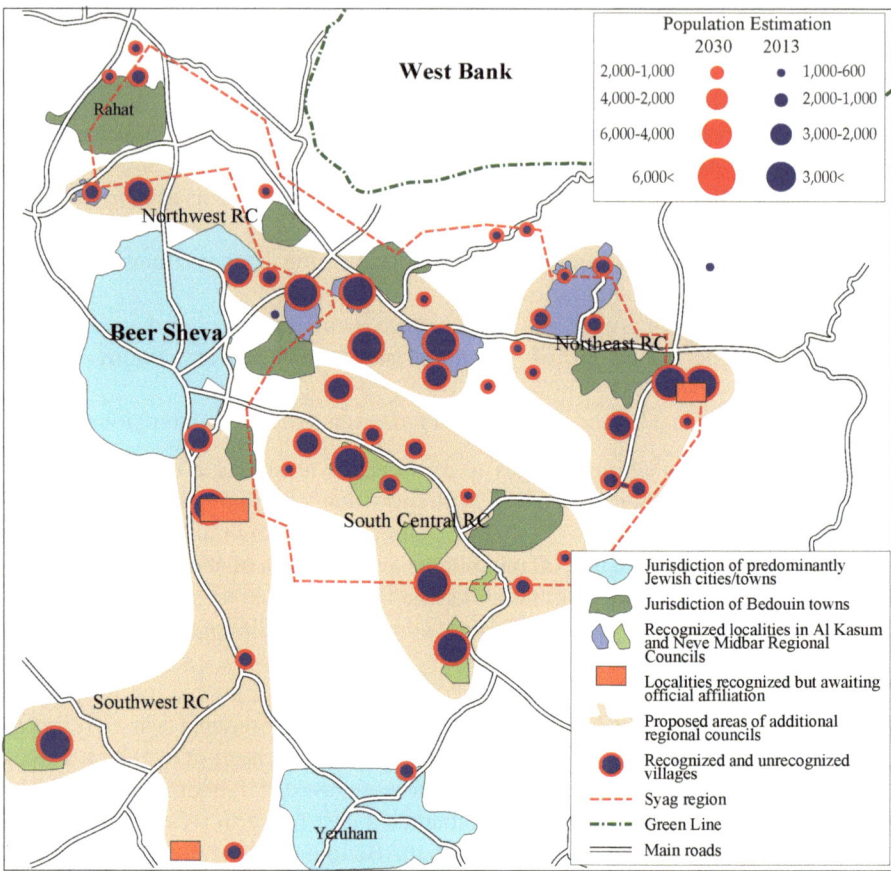

Fig. 9.8 Territorial distribution of proposed regional councils

east and the northwest. A northwestern regional council would be oriented toward Beer Sheva and the Bnai Shimon Regional Council, the northeastern one toward Arad, the south central toward Dimona, and southwestern council toward Yeruham. The councils would reflect the present clustering of villages along major tribal lines: the Azazme in the south, the Tiaaha in the east and north east, the Adhulam in the far east, and the Tarabin in the west. In drawing the boundaries for new regional councils, not only would tribal affiliation be important but so would the distribution of landless residents. Such a geographical expansion highlights the fact that the Syag's original boundaries have long been erased (for examples see Figs. 7.2, 7.3, and 9.8).

The most difficult negotiations will take place over the question of which unauthorized settlements should be recognized in situ and which should not. The latter would have to be relocated.

The paradigm presented in this book applies conflict management techniques to disputes over land claims and usage that arise from similar clashes between modernization and traditionalism. In presenting a model that offers planning options, the focus is not on the impasse between government proponents of concentrating the Bedouin and Bedouin tribal leaders who seek to maintain dispersed villages and extensive land areas. We have shifted the discourse to planning strategies that may enable the Bedouin to maintain as much as possible of their present lifestyles in suburban and exurban settings through regional council structures, or retain their communal identification through being annexed as kindred groups to adjoining town neighborhoods—both Arab and Jewish.

In these two options cultural clashes over such issues as education and job training for women and rigid leadership versus democratic political structures will inevitably affect negotiations. However, the interactive nature of the discourse should moderate these clashes and lead to compromises.

The strategy employed in the negotiations will have to take into account the absence of a unified approach by all stakeholders, including the Bedouin. This presents obstacles to Bedouin acceptance of shared regional services and infrastructure such as sewage, water, environmental protection, medical, educational, recreational and shopping facilities, and employment parks—and their resistance to any unified system of representation for purposes of land claims. The negotiations will have to be done case by case with each of the localities. Each case involves land issues over ownership (Meir 2009; Amara et al. 2012), jurisdictional claims, and planning/zoning rules (Yiftachel et al. 2012, updated 2014). Separation of jurisdictional claims and planning/zoning rules from ownership issues are critical for implementation of the Goldberg recommendations. The former must be dealt with immediately, while ownership claims can continue in the courts long after jurisdictional and planning recognition are negotiated. The negotiations table must include representatives of government ministries, economic agencies, village leaders, national and local authorities, and NGOs (e.g., RCUV, Civil Rights Union).

A major challenge for those coordinating the process is to implement the Goldberg Committee conclusions and reject efforts to significantly alter them. Otherwise, the report will become simply another addition to the history of failed efforts to reconcile the needs of the Negev Bedouin with the government's national goals.

The arena for negotiations that is proposed responds to the competing interests and needs of the differing stakeholders, rather than presenting the Bedouin and the authorities with fixed plans. Implementing the Goldberg Commission's recommendations represents both a challenge and a historic opportunity, and planners have a key role to play in its implementation. It is our hope that out of open discourse and mutual understanding, forms of government structures that respond to the inevitability of urban forces can be established that respect most of this minority's interests.

Conclusions

Resettlement policies for the Negev Bedouin must be revised, not through enforcement by the state, but developed in collaboration with the community. A framework for these negotiations should be initiated with both government representatives and representatives of the community in all its diversity who have decision-making powers and stature. For the government this means abandoning a top-down approach for a collaborative one. For the Bedouin leadership this means uniting internally to voice the varied needs of the population: land claimers, landless, rural, urban, older, and younger generations, to produce equitable economic, educational, and cultural desires and infrastructure requirements.

There needs to be a diversity of municipal frameworks: towns, varied types of rural villages, and integration for those who desire it in the neighborhoods of Beer Sheva. Solutions will range from those in the unrecognized villages filling in open lands (infill) in the towns and recognized villages based on geography and clan affiliation, there may be new regional councils, and others may integrate into the existing councils—Neve Midbar, Al Kasum, and the Jewish councils in the region. Currently the desire of both the Bedouin and Jewish Negev populations is for juris-dictional segregation. However examples should be taken from the north (the Gilboa and Misgav regional councils for example) where the Bedouin villages became part of already existing Jewish councils with recognition. Villages reaching a certain size may have independent municipal status. Not all villages will be recognized in situ—there are existing and planned national infrastructure which conflict. Land transfers en bloc or just compensation should be part of the negotiation agenda. The key is diversity. Just as the Jewish population has assorted lifestyle options, so too must the Bedouin. Indeed, as was discussed in Chap. 3, this lack of options applies to Israel's Arab minority as a whole, but much more extreme when the future of 55,700 Negev Bedouin is at stake.

The Bedouin have been a historical presence, rooted in the land for generations and loyal citizens of Israel. In terms of settlement policies it is imperative that the government and Bedouin collaborate to develop and implement plans that meet the diverse needs of the Bedouin community in a spirit of equity and justice while sat-isfying more general government interests. Shaping alternative land management and spatial planning processes which recognize the Bedouin right to a variety of urban and rural settlement options is crucial for rebuilding a positive relationship between the Bedouin and the State of Israel.

References

Abu Ras, T. (2006). Land disputes in Israel: The case of the Bedouin in the Negev, electronic version of Adala no. 24, April, quoting Ariel Sharon, land as an instrument to establish economic infrastructure and a significant reduction in social gaps. *Land, Bulletin of the Institute for the Study of Land and Land Use, 50,* 14–15. November–December.

Abu Sada, J., & Thawaba, S. (2011). Multi criteria analysis for locating sustainable suburban centers: A case study from Ramallah Governorate, Palestine. *Cities, 28,* 381–393.

Abu-Lughod, J. (1996). Urbanization in the Arab World and International System. In J. Gugler (Ed.), *Translated from the urban transformation of the developing world* (pp. 185–208). Oxford: Oxford University Press.

Abu-Rabia, K. (2012). *Tribes and political parties among the Negev Bedouin.* Beer Sheva: Negev Bedouin (Hebrew).

Abu-Rabia-Queder, S. (2006). Between tradition and modernization: Understanding Bedouin female dropout. *British Journal of Sociology of Education, 27*(1), 1–17.

Abu-Saad, K., Horowitz, T., & Abu-Sasd, I. (Eds.). (2007). *Weaving tradition and modernity: Bedouin women in higher education.* Beer Sheva: The Center for Bedouin Studies and Development and the Center for Regional Development, Ben Gurion University of the Negev.

Akbar, J. A. (1988). *Crisis in the built environment: The case of the Muslim City.* Leiden, ND: Concept Media Pte Ltd.

Al Hathlol, S. B. A. (1994). *The Arab Islamic City: impact of the legal in shaping the building environment.* Al-Reyde: Fahed Publications (Arabic).

Al-Dbag, M. (1988). *Beladona Falestine (Palestine Our Country).* Kafar Qara: Dar Elshafak (Arabic).

Al-Haj, M. (1995). Kinship and modernization in developing societies: The emergence of instrumentalized kinship. *Journal of Comparative Family Studies, 26*(3), 311–328.

Al-Haj, M., & Rosenfeld, E. (1990). *Arab local government.* Givat Haviva: Peace Studies Institute (Hebrew).

Altman, G, & Arbel, A. (2010). *Bedouin encroachment in the Negev: The Judge Goldberg report from a practical point of view.* The Institute for Zionist Strategies. Retrieved October 17, 2010, from www.izs.org.il/eng.

Amara, A., Abu-Saad, I., & Yiftachel, O. (Eds.). (2012). *Indigenous (In), justice; human rights law and Bedouin Arabs in the Naqab\Negev.* Cambridge, MA: Harvard University Press.

Amiad, N., & Moshkin, B. (Eds.). (2002). *Land: issues in land policy.* Jerusalem: Land Policy and Land Use Research Institute (Hebrew).

Appelbaum, L., & Hazan, A. (2005). *Cooperation between small local authorities: Lessons for Israel.* Jerusalem: Floersheimer Institute for Policy Studies (Hebrew).

© Springer International Publishing Switzerland 2015   91
D. Shmueli, R. Khamaisi, *Israel's Invisible Negev Bedouin,*
SpringerBriefs in Geography, DOI 10.1007/978-3-319-16820-3

Atzmon, E. (2013). The policy for settlement of the Bedouin in the Negev. In R. Bdhatzor (Ed.), *The Bedouin in the Negev: A strategic challenge to Israel* (pp. 50–63). S. Daniel Abraham Center for Strategic Dialogue, Natanya Academic College: Natanya.

Barlow, M., & Wastl-Walter, D. (Eds.). (2004). *New challenges in local and regional administration*. Surrey: Ashgate.

Ben David Y (1983) *The spontaneous development stages of the Negev Bedouin: From semi-nomadism to permanent settlement*. Doctoral Thesis, Hebrew University, Jerusalem (Hebrew).

Ben-David, Y. (1996). *Feud in the Negev: Bedouin, Jews, Land*. Rananna: The Center for the Research of Arab Society in Israel (Hebrew).

Ben-Elia, N. (2004). *The fourth generation: New local government in Israel*. Jerusalem: Floersheimer Institute for Policy Studies (Hebrew).

Ben-Elia, N. (2006). *From municipal to supra-municipal modes of service delivery in Israel*. Jerusalem: Floersheimer Institute for Policy Studies. Pub. 1/57 (Hebrew).

Berke, P. R., Ericksen, N., Crawford, J., & Dixon, J. (2002). Planning and indigenous people: Human rights and environmental protection in New Zealand. *Journal of Planning Education and Research, 22*(2), 115–134.

Berry, B. J. L. (Ed.). (1976). *Urbanization and counterurbanization*. Beverly Hills, CA: Sage.

Bourne, L. (2001). The urban sprawl debate: Myths, realities and hidden agendas. *Plan Canada, 41*, 26–30.

Braverman, I., Blomley, N., Delaney, D., & Kedar, A. (Eds.). (2014). *The expanding spaces of law: A timely legal geography*. Standard, CA: Stanford University Press.

Cabinet Resolution No. 1999, *Establishment of an authority for Bedouin settlement arrangements in the Negev, 15 July 2007*. Unofficial Translation.

Cannana, T. (1933). *The Palestine Arab House: its architecture and folklore*. Jerusalem: The Syrian Orphanage Press.

Central Bureau of Statistics. (1949). *Israel statistical yearbook, No. 1*. Jerusalem: Central Bureau of Statistics.

Central Bureau of Statistics (CBS). (2014). *Statistical abstract of Israel, No. 65*. Jerusalem: Central Bureau of Statistics.

Centre on Housing Rights and Evictions (COHRE). (2008, February). Report submission to the Goldberg Committee regarding violations of the human right to water and sanitation in the unrecognized villages of the Negev/Naqab.

Champion, T., & Hugo, G. (Eds.). (2004). *New forms of urbanization: Beyond the urban rural dichotomy*. Ashgate: Aldershot Hants.

Consensus Building Institute. (2006). *Conflict assessment report: Development disputes in Kaseife and Um Batin (An analysis of conflicts between the Government of Israel and Bedouin stakeholders in the Negev Desert Region of Israel)*. Cambridge, MA: The Consensus Building Institute.

Dale, N. (1999). Negotiating the Future of the Haida Gwaii. In L. Susskind, S. McKearnan, & J. Thomas-Larmer (Eds.), *The consensus building handbook* (pp. 923–950). Thousand Oaks, CA: Sage.

Daniel, E. J., & Kalchheim, C. (Eds.). (1988). *Local government in Israel*. Lanham, MD: Jerusalem Center for Public Affairs and University Press of America.

Dayan, A. (2010). *Kindergarten Conflict, Jerusalem report*. Retrieved August 14, 2010, from www.jpost.com/JerusalemReport/Article.aspx?id=184712.

Dollery, B., Burns, S., & Johnson, A. (2005). *Structural reform in Australian local government: The Armidale Dumaresq-Guyra-Uralla-Walcha Strategic Alliance Model*, working paper. University of New England, Retrieved from www.une.edu.au/bepp/working-papers/economics/1999-2007/econ-2005-2.pdf.

Dor-On, A. (Ed.). (2009). *Implementation of the Goldberg Committee Recommendations – Dangers and opportunities, land: Issues in land policy*. Jerusalem: Land policy and Land Use Research Institute. No. 66, May (Hebrew).

Eben Saleh, M. A. (2002). The transformation of residential neighborhood: The emergence of new urbanism in Saudi Arabian culture. *Building and Environment, 37*, 515–529.

Elazar J. D., & Kalchheim, C., (Eds.), (1988). *Local Government In Israel*, Lanham, Md: University Press of America.

Efrati, Y., Razin, E., & Brender, E. (2004). *Reform in local government: decentralization for the worthy and accessorization for the underprivileged*. Jerusalem: Israel Democracy Institute (Hebrew).

Elshastawy, Y. (2004). *Planning Middle Eastern Cities: An Urban Kaleidoscope in a Globalizing World*. London: Routledge.

Elshesawy, Y. (Ed.). (2008). *The evolving Arab City: Tradition, modernity and urban development*. London: Routledge.

El-Taji, M. (2007). Arab local councils in Israel: the democratic state and the Hamula. In F. Lazin, H. Wollmann, V. Hoffmann-Martinot, & M. Evans (Eds.), *Local government reforms in countries in transition: A global perspective* (pp. 253–272). London: Lexington Books.

El-Taji, M. (2008). *Arab local authority in Israel: Hamulas, nationalism and dilemmas of social change. Doctorate Dissertation*. Seattle, WA: University of Washington.

Falah, G. (1989). *Al-Phelstenyon Almnseyon- Arab Alnaqap 1906–1989 (The forgotten Palestinians – Arabs in the Negev 1906–1089)*. Beirut: Institute for Palestinian Studies (Arabic).

Goldberg Commission. (2008). Retrieved January 10, 2010, from www.moch.gov.il/NR/rdonlyres/770ABFE7-868D-4385-BE9A-96CE4323DD72/5052/DochVaadaShofetGoldbergHebrew3.pdf.

Hasson, S. (2006). *The local democracy deficit* (pp. 17–27). Tel Aviv: Proceedings of the Second Conference on Local Government, Tel Aviv University (Hebrew).

Healey, P. (1997). *Collaborative planning*. London: McMillan.

Henderson, J. V., & Wang, G. H. (2005). Aspects of the rural-urban transformation of countries. *Journal of Economic Geography, 5*, 23–43. Retrieved from http://joeg.oxfordjournals.org/content/5/1/23.full.pdf+html.

Hezmawe. (1998). *Land ownership in Palestine, 1918–1948*. Akka: Al-Aswar Institution. Publisher (in Arabic).

Hibbard, M., Lane, M. B., & Rasmussen, K. (2008). The split personality of planning for land and resource management. *Journal of Planning Literature, 23*(2), 136–151.

Hidalgo, M. A., & Hernandez, B. (2001). Place attachment: Conceptual and empirical questions. *Journal of Environmental Psychology, 21*(3), 273–281.

Hlihel, A. (2011). Barriers and motives for internal immigration among Palestinian citizens of Israel. In R. Khamaisi (Ed.), *The Arab's society in Israel (4), population, society, and economy* (pp. 69–86). Tel-Aviv: Van Leer Institute and Hakibbutz Hameuchad Publishing House (Hebrew).

Howard, B. R. (2003). *Indigenous peoples and the state: The struggle for native rights*. DeKalb, IL: Northern Illinois University Press.

Huntington, S. P. (1996). *The clash of civilizations and the remaking of world order*. New York, NY: Simon & Schuster.

Innes, J. (1995). Planning theory's emerging paradigm: Communicative action and interactive practice. *Journal of Planning Education and Research (Spring), 14*(3), 183–189.

Innes, J. (1996). Planning through consensus building: A new view of the comprehensive planning ideal. *Journal of the American Planning Association (Autumn), 62*(4), 460–472.

Innes, J., & Booher, D. (1999). Consensus building as role playing and bricolage. *Journal of the American Planning Association (Winter), 65*(1), 9–26.

Kafkoula K (2009) New towns, international encyclopedia of human geography, pp. 428–437.

Kedar, A. (2001). Legal transformation of ethnic geography: Israeli law and the Palestinian landholder 1948–1967. *Journal of International Law and Politics, 33*(4), 923–1000.

Kedar, S., & Yiftachel, O. (2006). Land regime and social relations in Israel. In H. de Soto & F. Cheneval (Eds.), *Realizing property rights: Swiss human rights book* (Vol. 1, pp. 129–146). Zurich: Ruffer and Rub Publishing House.

Khamaisi, R. (1990). *Master plan for Bedouin settlement*. Beer Sheva: Association for the Assistance and Protection of the Bedouin Population (Hebrew).

Khamaisi, R. (1996). *New towns alongside existing towns*. Jerusalem: The Florshimer Institute for Policy Studies.

Khamaisi, R. (2002). *Toward the expansion of the areas of jurisdiction of Arab Communities in Israel*. Jerusalem: Floersheimer Institute for Policy Studies (Hebrew).

Khamaisi, R. (2004). Urbanization without cities: The urban phenomena among the Arabs in Israel. In: Maos JO, Inbar M, Shmueli DF (eds) *Contemporary Israeli geography. Horizons in geography, 60–61*, 41–50.

Khamaisi, R. (2005). Between Town and Village: continuity and change in Arab localities in Israel. In M. Al-Haj, M. Saltman, & Z. Sable (Eds.), *Social critique and commitment* (pp. 107–122). New York, NY: University Press of America.

Khamaisi, R. (2008). Arab local authorities: A transient or a structural crisis? In A. Mana'a (Ed.), *The book of Arab Society in Israel 2* (pp. 409–438). Jerusalem: Van Leer Institute and Hakibbutz Hameuchad (Hebrew).

Khamaisi, R. (2009). Think Out of the Box, *Karka*, 66, 69–85. (Hebrew).

Khamaisi, R. (2010). Between the Hammer and the Anvil: Spatial and structural barriers in outline planning of the Arab Localities in Israel. In A. Haidar (Ed.), *The collapse of Arab local authorities: Suggestions for restructuring* (pp. 47–75). Tel Aviv: Van Leer Jerusalem Institute/ Hakibbutz Hameuchad Publication House (Hebrew).

Khamaisi, R. (2012). Transition from ruralism to urbanization: The case of Arab localities in Israel. *Horizons in Geography, 80*(Special Issue: Themes in Israeli Geography), 168–183.

Khamaisi, R. (2013). Barriers to Developing Employment Zones in the Arab Palestinian Localities in Israel and Their Implications. In N. Khattaab & S. Miaari (Eds.), *Palestinians in the Israeli Labour Market: A multi-disciplinary approach* (pp. 185–212). New York, NY: Palgrave Macmillan.

Khamaisi, R., & Shmueli, D. (2009). *Abu Basma Regional Council: Preparing for municipal reorganization (final report for the Abu Basma Regional Council)*. Haifa: University of Haifa (Hebrew).

Khattaab, N., & Miaari, S. (2013). *Palestinians in the Israeli Labour Market: A multi-disciplinary approach*. New York, NY: Palgrave Macmillan.

Khoury J, Yagna Y (2010) Police destroy dozens of buildings in unrecognized Bedouin village in Negev, HaAretz. Retrieved July 28, 2010, from www.haaretz.com.

Kipnis, B. (1976). Trends of the minority population in the Galilee and their planning implications. *Ir ve'Ezor* [City and Region], *3*, 54–68 (Hebrew).

Koeller, K. (2006). The Bedouin of the Negev: A forgotten minority. *Forced Migration Review, 26*, 38–39.

Kretzmer, D. (2002). *The legal status of the Arabs in Israel*. Jerusalem: The Institute for Israeli Arab Studies and the Van Leer Institute. Updated edition (Hebrew).

Lane, M. B. (2001). *Indigenous land and community security: A radical planning agenda Working paper no 45*. Madison, WI: Land Tenure Center, University of Wisconsin-Madison.

Lane, M. B. (2002). Buying back and caring for country: Institutional arrangements and possibilities for indigenous lands management in Australia. *Society and Natural Resources, 15*, 827–846.

Lane, M. B. (2005). The role of planning in achieving indigenous land justice and community goals. *Land Use Policy, 23*(4), 385–394.

Lane, M. B., & Hibbard, M. (2005). Doing it for themselves: Transformative planning by indigenous peoples. *Journal of Planning Education and Research, 25*, 172–184.

Lazin, F., Wollmann, H., Hoffmann-Martinot, V., & Evans, M. (Eds.). (2007). *Local government reforms in countries in transition: A global perspective*. Lanham, MD: Lexington Books.

Lithwick, H. (2002). *Making Bedouin towns work*. Jerusalem: The Center for Social Policy Studies in Israel.

Mai, M. M., & Shamsuddin, S. (2008). Ethnic spatial identity in the context of urbanization: The transformation of Gbagyi compounds in North Central Nigeria. *Journal of Urbanism, 1*(3), 265–280.

Marx, E. (1980). Wage, Labor and Tribal Economy of the Bedouin in South Sinai. In P. Zalzman (Ed.), *When nomads settle* (pp. 111–123). New York, NY: Praeger.

Marx, E. (2005). Nomads and cities: The development of a conception. In S. Leder & B. Streck (Eds.), *Shifts and drifts in nomadic sedentary relations*. Wiesbaden, GE: Nomaden and Sesshafte 2.

Masry-Herzalla, A., Razin, E., & Choshen, M. (2011). *Jerusalem as an internal migration destination for Israeli-Palestinian families*. Jerusalem: Floersheimer Institute for Policy Studies (Hebrew).

Meir, A. (1997). *As nomadism ends: The Israeli Bedouin of the Negev*. Boulder, CO: Colombia.

Meir, A. (1999). Local Government among Marginalized Ex-nomads: The Israeli Bedouin and the State. In H. Jussila, R. Majoral, & C. C. Mutambirwa (Eds.), *Marginality in space-past present and future* (pp. 101–119). Aldershot, UK: Ashgate.

Meir, A. (2003). *From planning advocacy to independent planning: The Negev Bedouin on the path to democratization in planning*. Beer Sheva: The Negev Center for Regional Development, Ben Gurion University of the Negev (Hebrew).

Meir, A. (2005). Bedouin, the Israeli state and insurgent planning: Globalization, localization or glocalization. *Cities, 22*(3), 201–215.

Meir, A. (2009). Contemporary state discourse and historical pastoral spatiality: Contradictions in the land conflict between the Israeli Bedouin and the State. *Ethnic and Racial Studies, 32*(5), 823–843.

Meir, A., & Marx, E. (2005). Land, towns and planning: The Negev Bedouin and the state of Israel. *Geography Research Forum, 25*, 43–62.

Meir-Brodnitz, M. (1986). *Social aspect in planning in the Arabs sector, the regulative planning and the process of self-housing*. Haifa: Center for City and Region Research, Technion (Hebrew).

Moravcsik, A. (2004). Is there a democratic deficit in world politics? A framework for analysis. *Government and Opposition, 39*(2), 336–363.

Nachmias, D. (2006). Governance of local government in Israel: Policy analysis and reformulation. In A. Reichman & D. Nachmias (Eds.), *The state of Israel: New thoughts* (pp. 83–113). Herziliya: Mifalot – Interdisciplinary Center (Hebrew).

Nasasra, M., Richter-Devroe, S., Abu-Rabia-Quedar, S., & Ratcliffe, R. (Eds.). (2014). *The Naqab Bedouin and colonialism: New perspectives*. New York, NY: Routledge Studies on the Arab-Israeli Conflict.

Negev Coexistence Forum for Civil Equality. (2009, August). The Bedouin-Arabs in the Negev-Naqab desert in Israel, *Shadow report response to the report of the state of Israel on implementing the covenant on civil and political rights (CCPR)*.

Ozacky-Lazar, S., & Ghanem, A. (Eds.). (2003). *The Orr Testimonies: Seven professional opinions submitted to the Orr Committee*. Jerusalem: Keter Publishing House Ltd.

Porat, H. (2009). *Bedouin in Negev: Between nomadic and urbanization*. Beer Sheva: Negev Center for Regional Development, Ben-Gurion University.

Pugh, C. (1995). *Urbanization in developing countries: An overview of economic and policy issues in the 1990s*. Sheffield: Sheffield Hallam University.

Razin, E. (2003a). *Reform in the organization of local government in Israel: Between centralization and decentralization, between traditionalism and modernity*. Jerusalem: Floersheimer Institute for Policy Studies (Hebrew).

Razin, E. (2003b). *Local government reform in Israel: Between centralization and decentralization, between traditionalism and modernity*. Jerusalem: Floersheimer Institute for Policy Studies (Hebrew).

Razin, E., & Hasson, S. (1992). Defining the boundaries between regional councils and development towns – Principles and lessons from the conflict concerning the regional projects in the Negev. *Ir Va-Ezor, 22*, 72–89 (Hebrew).

Razin, E., Hasson, S., & Hazan, A. (1994). The struggle over municipal boundaries: Regional councils and urban space. *Ir Va-Ezor, 23*, 5–28 (Hebrew).

Sandercock, L. (Ed.). (1998). *Making the Invisible visible: A multicultural planning history*. Berkeley, CA: University of California Press.

Sanders, D. (1999). Indigenous peoples: Issues of definition. *International Journal of Cultural Property, 8*, 4–13.

Shemony, Y. (1949). *Arabs of Eeretz Yezrael* [Land of Israel], Tel Aviv: Am-Oved Publishers (Hebrew).

Shmueli, A. (1980). *The end of nomadism: Patterns of settlement in Bedouin societies.* Tel Aviv: Reshafim (Hebrew).

Shmueli, D. (2005). Is Israel ready for participatory planning? Expectations and obstacles. *Planning Theory and Practice, 6*(4), 485–514.

Shmueli, D. (2008). Environmental justice in the Israeli context. *Environment and Planning A, 40*, 2384–2401.

Shmueli, D., & Khamaisi, R. (2011). Bedouin communities in the Negev: Models for planning the unplanned. *Journal of the American Planning Association, 77*(2), 109–125.

Shmueli, D., Warfield, W., & Kaufman, S. (2009). Enhancing community leadership negotiation skills to build civic capacity. *Negotiation Journal, 25*(2), 249–266.

Soffer, A. (2009). The Negev as a strategic asset will be transformed into an abandoned region – Israel will become the State of Tel Aviv. In A. Dor-On (Ed.), *Implementation of the Goldberg Committee Recommendations – Dangers and opportunities, land: issues in land policy* (Vol. 66, pp. 17–35). Jerusalem: Land Policy and Land Use Research Institute (Hebrew).

Song, Y., Zenou, Y., & Ding, C. (2008). Let's not throw the baby out with the bath water: The role of urban villages in housing rural migration in China. *Urban Studies, 45*(2), 313–330.

Stern, E., & Gradus, Y. (1979). Socio-cultural considerations in planning towns for nomads. *Ekistics, 227*, 224–230.

Sternberg Ben-Baruch, D. (2013). Planning of Bedouin Communities after recognition in comparison with planning processes in Arab and Jewish communities – Case studies from the Galiliee. *Ph.D. Thesis*, Department of Geography and Environmental Studies, University of Haifa, Haifa.

Susskind, L., & Auguelovs, I. (2008). *Addressing the land claims of indigenous peoples. Program on human rights and justice.* Cambridge, MA: MIT Center for International Studies (CIS).

Susskind, L., & Cruikshank, J. (1987). *Breaking the Impasse: Consensual approaches to resolving public disputes.* New York, NY: Basic Books.

Susskind, L., & Field, P. (1996). *Dealing with an angry public: The mutual gains approach to resolving disputes.* New York, NY: The Free Press.

Susskind, L., McKearnan, S., & Thomas-Larmer, J. (1999). *The consensus building handbook.* Thousand Oaks, CA: Sage.

Tarrabeih, H., Khamaisi, R., & Shmueli, D. (2012). Intensification of environmental conflicts resulting from rural to urban transformation: The case of Arab localities in a changing Galilee. *Geographical Research Forum, 32*, 5–27.

United Nations. (2008). *World urbanization prospects: The 2007 revision.* New York, NY: Department of Economic and Social Affairs. http://www.un.org/esa/population/publications/wup2007/2007WUP_Highlights_web.pdf.

Wang, Y. P., Wang, Y., & Wu, J. (2009). Urbanization and informal development in China: Urban villages in Shenzhen. *International Journal of Urban and Regional Research, 3*(4), 957–973.

Yagna, Y. (2010). *Israel to triple demolition rate for illegal Bedouin Construction, HaAretz.* Retrieved February 18, 2010, from www.haaretz.com/hasen/spages/1150664.html.

Yahel, H., Kark, R., & Frantzman, J. S. (2013). Are the Negev Bedouin an indigenous people? *Ofakim in Geography, 84*, 73–87 (in Hebrew).

Yahodkin, S. (2004). *The unrecognized villages: Recognition and equal rights (position paper).* Tel Aviv: The Regional Council for the Unrecognized Villages in the Negev, Bimkom and The Arab Center for Alternative Planning (Hebrew).

Yiftachel, O. (1992). *Planning a mixed region in Israel.* Aldershot: Aldershot Publishing.

Yiftachel, O. (1999). Between nation and state: 'Fractured' regionalism among Palestinian-Arabs in Israel. *Political Geography, 18*(3), 285–307.

Yiftachel, O. (2001). Introduction: Outlining the Power of Planning. In O. Yiftachel, J. Little, D. Hedgecock, & I. Alexander (Eds.), *The power of planning: Spaces of control and transformation* (pp. 1–20). Dordrecht, GE: Kluwer.

Yiftachel, O. (2002). Ethnocracy: The politics of Judaizing Israel/Palestine. Retrieved January 13, 2015, from http://www.geog.bgu.ac.il/members/yiftachel/new_papers_eng/Constellations-print.htm.

Yiftachel, O. (2006). Inappropriate and unjust: Planning for private farms in the Naqab [Negev]. *Adalah's Newsletter, 24*, 7 pages.

Yiftachel, O. (2009). Theoretical notes on 'Gray Space': Mobilization of the colonized. *City, 13*, 240–256.

Yiftachel, O., Baruch, N., AbuSammur, S., Sheer, N., & BenArie, R. (2012). *A master plan for the unrecognized Bedouin villages in the Negev, selected sections*. Jerusalem: The Regional Council for the Unrecognized Villages in the Negev (RCUV) and BIMKOM Planners for Planning Rights. Updated 2014.

Zaferatos, N. C. (1998). Planning and the Native American tribal community: Understanding the basis of power controlling the reservation territory. *Journal of the American Planning Association, 64*(4), 395–410.

Zaferatos, N. C. (2004). Tribal nations, local governments, and regional pluralism in Washington State: The Swinomish approach in the Skagit Valley. *Journal of the American Planning Association, 70*(1), 81–96.

Zandberg, H. (2009). Land ownership dispute is not the central issue in the "Bedouin Problem". In A. Dor-On (Ed.), *Implementation of the Goldberg Committee Recommendations – Dangers and opportunities, land: Issues in land policy, No. 66*. Jerusalem: Land policy and Land Use Research Institute (Hebrew).

Index

© Springer International Publishing Switzerland 2015
D.F. Shmueli, R. Khamaisi, *Israel's Invisible Negev Bedouin*,
SpringerBriefs in Geography, DOI 10.1007/978-3-319-16820-3